WE KNOW NOT WHAT WE DO

Based On the Award Winning Documentary

ELIZABETH GAYLYNN BAKER

Acknowledgments

Oh so many, but I must begin with Sanjay Burman of Burman Books who from the beginning has felt like a soulmate in consciousness. You have accepted me, inspired me, believed in me, always encouraged me, and now have published me twice…*Gifts of Gratitude*. (www.giftsofgratitude.net) and *We Know Not What We Do* (www.weknownotwhatwedo.net) Thank you doesn't begin to cover how magical our relationship continues to be Sanjay. While I am at it, let me give special thanks to Gemma Wood, my great editor on this one. You taught me a lot, and have genuinely helped me find my voice.

There is a gathering of magical artists giving their best to Mother Earth. Doron Gazit and Duby Tal are two of them. I send them heartfelt appreciation for creating the deeply mysterious creature on the cover. Traveling around the world with his 'red line project, Doron (www.dorongazit.com) 'draws' with three dimensional lines, sculpting the air, reflecting the sun, and creating a dialogue with nature to demonstrate environmental degradation. His bright red lines, in bleak landscapes, create haunting images alerting observers to the urgent need to remedy and protect our endangered environment. Duby Tal's (www.albatross.co.il) amazing camera skill, captures our planet in the air, on the ground, and straight through the heart. This particular cover picture was chosen to signify that the profound mystery of our Planet defies description.

I have learned that life itself writes books, so I must thank Aaron Daniel Taylor and his lovely wife, Rhiannon Taylor, who began this adventure by hiring me to make their film, *We Know Not What We Do*, and faithfully supported my vision during our journey.

And in no certain order, I want to acknowledge and thank Grant Taylor, Matt Schulze, Michael Clark, Barbara and Alvaro Cardonna-Hine, HawaH, Hakim Bellamy, Jerry Fairies, Ramon Bermudez Jr, Phoenix Avalon, Swang Lin, Amy Faires, Eric Faires, Madi Sato, Caroline Herring, Jane Barnham, Price Hall, Patrick and Faye Bates, and Eric Zerkel, and all the incredibly ingenious artists who supported me while the film and the book was create.

A very special thanks to Shirley MacLaine and Brit Elders, extraordinary souls in my life, as well as Steven and Deirdre Evans, Debra Fuller, Tegi and Bruce MacGregor, Adrianna and Cher Tredanari, Hollace Davids, Chevonne 0'Shaughnessy, Mary Leary, Tom Donnelly, Sandy Lefler, Rosebud Reservation, and even my horse love, Mine That Bird who continues to let me ride him in our shared dreams. Thank you all for your love, support, and inspiration.

With deep gratitude, I appreciate my teachers, on both sides of the veil, who continue to shape my life, and my two newest teachers, Chief Leonard Crow Dog, and wounded healer, Paul Levy who answers so many unanswered questions in his spellbinding book, *Dispelling Wetiko…Breaking the Curse of Evil*.

Finally of course, with all my love, my family; the pure joy of my grandchildren, Kira, Zoe and Zen, and my sons and their mates, all of whom can never be appreciated enough. It is simply life's very best adventure to be their Yuppa Du.

Elizabeth Gaylynn Baker
goodfilms11@gmail.com
July 20, 2016

"I wish that books like "We Know Not What We Do" didn't have to exist and that Elizabeth Gaylynn Baker was writing another heart-warming story to feed our souls. I wish that all the warning signs on climate change and the man-made environmental catastrophe that Elizabeth addresses in this marvelous book were being heeded. However, wishing is not enough. "We Know What We Do" is the wake-up call, indeed the clarion call to action. It shows us that we must all act, and act now, and that we must learn to 'know what we are doing' before it is to late...

– Thomas John Donnelly. MA. BA.
Associate Principal Lecturer Media, Film and Culture

Stories about climate change, based on scientific evidence, no matter how compelling, do not deeply move people. But when the moral bone is hit, the call to action is powerful. I'm amazed and grateful and moved to tears.

– Barbara Cardona-Hine
Artist

"We Know Not What We Do is a powerful testament to the human spirit. This is a supremely crafted story of hope and resilience woven together with heart and courage. I was honored to be a part of this production and such accomplished and important individuals who are shaping the world."

– Hawah Kasat
Artist, Author, Community Organizer
Cofounder: One Common Unity

DEDICATION

I am dedicating this collection of interviews to the brave men and women
from all walks of life, from all their different spiritual paths, who took the
time to sit with me and tell me their stories. Their explanations of how
they are spending their lives, caring for others as they care for Planet Earth,
both inspired and then changed and deepened my own spiritual life.
And for that I am very grateful.

Elizabeth Gaylynn Baker

Contents

*In our hunger for wealth and power, we impoverish our home,
the Earth that sustains us. We think there will be
no consequences...but we are wrong...*

WE KNOW NOT
WHAT WE DO

LOVE STORIES... AN INTRODUCTION

By Rev. Richard Cizik
Center for the New Evangelical Partnership
www.NewEvangelicalPartnership.org

There's a beautiful expression, "The world is a story." Most human beings do not believe this, or perhaps only see their lives, not the world, this way. Alas, I call this deficit what it is -- postmodernism -- since to live without stories, narratives, or symbols is to be divided, and to sadly face the future without the sense of community. A position that makes us vulnerable to exploitation by false storytellers.

I personally believe the most universal story that crosses all religions, cultures, political boundaries, and even time, is our identity as stewards of the Earth. It binds us together, and makes it impossible to see that we are all just solitary inhabitants, with no connection to one another.

In the compelling and provocative film "We Know Not What We Do," Elizabeth Baker has assembled a cast of storytellers; each with a voice of passion, telling us the Earth is in great peril. But it's not their concern that speaks as much as it is their hearts, their connection to the Earth, and love for it. That love is palpable, and even comes through in addressing what we all regard to be a crisis unlike anything prior in the history of the world.

Crisis is the Greek word for judgment. Theologians instruct us that judgment is best understood, not as God peevishly casting down what needs admonishment, but that moment of truth when what is sown is that which is reaped in a divinely ordered world.

People of many different faiths, with many different perspectives, are united because we have reached such a time in the history of the planet, in the 'story' of our world. Their voices are filled with love, clear, and unmistakable. We see them in the film, and now can read them speaking truths about the strip-mining in the Appalachian mountains, the heartland of America being threatened by the KXL Pipeline, and even, in my estimation, the legalized bribery of our politicians and public officials by greedy, exploitative, representatives of the fossil fuel industry. For in him all things were created: things in heaven and on earth, visible and invisible, whether thrones or powers or rulers or authorities;all things have been created through him and for him (Colossians 1:16).

But to adequately respond to this crisis, we must be able to see and think more clearly; care more deeply, and act more boldly. We must make a total paradigm shift. It requires more than simply going on with life as usual, albeit it with more enthusiasm. It is a move from an anthropocentric to a cosmocentric worldview. ("For God so loved the cosmos...")

The strength of the storytellers in the film and accompanying book are that they get this. They state the moral imperative and explain what it means, with the verbal insights and capacity to communicate it (something scientists and policy makers have been unable to do), and then translate for the viewer and reader that new vision into an ethical framework of what to do, and what needs to be done.

It is my privilege to introduce to you each of these powerful voices. I join them in a belief that we can change hearts and minds. And none too soon. I believe that the climate of opinion on climate change needs to change more rapidly than the climate itself, if we are to avoid the disasters described in this book and depicted in "We Know Not What We Do."

If we are successful, if we can do this, one result will be the de-politicizing of the climate change debate. Imagine that? It's called "taking impossible challenges and making them inevitable." It is our capacity as creatures of the Creator to be about our highest and best calling, call it creation care, call it sustainability, call it stewardship, call it what you want -- it's what the Creator intended.

My prayer is that this book will disturb you with the truth of what's happening to our planet, unsettle you with deepened conviction that we must act, and then lift up your spirit with hope. It can be done. It is being done. Thy will be done on Earth, as it is in heaven.

Some of my favorite words are those of John Wesley, a Methodist evangelist and early environmentalist in our American history, who warned his followers not to go into the mines, lest they die from coal dust:

"To serve the present generation, my calling to fulfill; Oh may it all my powers engage, to do my Master's will; Arm me with jealous care, as in Thy sight to live; and oh, thy servant lead, prepare a strict account to give."

Prologue:

"What the heck do I know about Climate Change?"

You may be asking the same thing of yourself. It's a question that many of us may be asking ourselves in these forthcoming days. The question rolled over and over in my mind as I thought about the offer that had just been made. First and foremost, the current offer was a job, and it was not just a normal job - but a job I would actually love. Aaron Daniel Taylor was offering me the job of making a documentary film.

Documentary or feature, filmmaking is great creative fun, but climate change? Good grief, who could understand or make a small indie film about a subject that LARGE? "Does it have to be about climate change?", I wailed to the empty room as I hung up the phone. Why couldn't it be about something I could get my arms around? Something easier? But 'easier' has never seemed to be on the name of the paths that find me on planet Earth and, who knows, maybe 'easier' doesn't even exist here. The blaring truth was, climate change or climate the same, I needed the job.

You can understand that awkward position these days, can't you? That situation called needing a job? I not only needed a job but, as I looked in all directions, this was the only job being offered. So even though I said it several times 'What the heck do I know about Climate Change?' I also said yes. What I didn't understand was that little 'yes' was about to change my life forever.

But wait a minute! What does this have to do with you? Why even bother to read a book if it doesn't have something to do with you? Is it going to help you get your next job? Or feed your family? Or help you have more success in your life? I sincerely hope the answers to these questions are yes! Because to read this classic story, of someone taking the hero's journey, of biting off more than they can chew, changes us, and in my opinion, changes us for the better. Here's how:

To follow along with me, and hear this particular tale of 'our' Earth, told out of so many different mouths, from a multitude of faiths, from many dissimilar backgrounds, and in such splendid voices, expands us. It altered me to have my 'stereotypes' blown apart. It transformed me to learn how foolish my choices have been, and to realize how many ways I've neglected my life-giving Planet. I am putting this book out into the world in the sincere hope that it will do the same for you.

Before this adventure was finished, my question would be more like what the heck did I actually know about anything? So please be patient, whatever your faith or beliefs - Christian, Native, Scientific or Other. Take my hand and let us make this mysterious journey together.

Shhhhh, the Planet is speaking.

Elizabeth Gaylynn Baker

It is Complicated...

I said to a squirrel, "What is that you are carrying?" And he said,
"It is my lucky rock; isn't it pretty?" I held it and said, "Indeed."
I said to God, "What is this earth?" And He said,
"It is my lucky rock; Isn't it wondrous?
And I said, "Yes - indeed."
16th Century Monk

On top of it being a difficult subject that was already eating like acid through the metal of my heart, my producer, Aaron Daniel Taylor, is an Evangelical Christian Conservative. He was actually asking me to make a film that would change the heart and minds of his fellow Evangelicals, among whom many climate deniers dwell. I didn't have the slightest idea of how to do that.

Aaron lined up the first interview in Appalachia with Peter Illyn of 'Restoring Eden'. Peter is a Christian Conservative Environmentalist and I was too green to the subject to realize what an unusual mixture that is. But I was excited that Appalachia was our first stop, and envisioned the Smokey Mountains that I had heard tales of all my life.

Peter is a bear of a man, with one eye. He wears a black eye patch that makes him look like a pirate. Refreshingly candid, with a wicked sense of humor, he is immediately endearing. Our first stop was the auditorium of a Christian College. Peter was presenting their final chapel service before graduation. I was expecting to see bright eager faces. What I saw instead was a bunch of bored sleepy students who had to be there in order to graduate. They all clutched their carry out coffee containers as beacons to help them find their seats through half closed eyes.

Peter had inspiring music, a slide show, and a real skill at story telling. Surely his words and pictures would arouse the audience. Wrong again. I wondered what I was missing. "Can't you hear him?" I wanted to shout! "Don't you understand what he is saying?" The message couldn't be clearer; 'the end is near unless we wake up and change our path'. The students could hardly hide their jaded yawns.

First came music, wonderful stuff with fiddles and a standing base, and then Peter began to speak, underlying his words with pictures from his slide show. As the images began to flash and Peter's deep rich baritone filled the space, an amazing thing happened to me. A huge whale rose up out of the ocean and aligned its tail with the top of a mountain. The unexpected vision only lasted seconds. Just long enough to burst open my chest, and surprise me with a flood of silent tears. The whale carries the wisdom of the ocean. The mountains store the wisdom of the land. We humans are destroying both. Visions come to touch and open our hearts. Thus began my apprenticeship. This was no longer 'just a job'. Unexpectedly, I now understood a great deal more about climate change than I had just seconds earlier.

Peter Illyn
Founder and Director, Restoring Eden
www.RestoringEden.org

"I was a Boy Scout growing up," Peter began. One night we took a hike in the mountains and I went to sleep on a bed of moss that lit up in the middle of the night. It was phosphorous, and I remember being awed. Today my Russian Orthodox background actually supports my environmental ministry. To the Orthodox, Jesus' sacrifice on the cross was to reconcile all of nature, all of creation, back to God. This idea, that the sacrifice of Jesus was to redeem all of creation, has become my springboard for environmental stewardship.

I lost my eye to cancer. I had melanoma. I don't believe it was the devil. I don't believe it was God. I believe it was too much sun on my retina. But once that cancer was there and it could spread, I had really no choice but to have my eye removed. Within the evangelical world there's this idea of right relationship with God, right relationship with my neighbor, and right relationship with Creation. With my heart I see faith not only as a destination but also as a journey. I think that's a very important part of who I am.

The '80's were a tumultuous time within the Evangelical Christian world. For the first time, with the rise of the political right, there was suddenly a divide between ideology and theology. There was an increasing belief that, if you were a Christian, you had to be a Republican, but I would look at the Republican agenda and see the deficits. Then I would examine the Democratic agenda and find deficits there too. Both were bringing something to the table, and so I began to struggle, as a centrist, in a world that was becoming increasingly polarized, and I really didn't know how to manage the tension.

I decided to take a sabbatical from church work, and do what I had always wanted to do: take a long walk in the woods. There was a famous book called 'My Side of the Mountain' that I had read as a kid. It was about a boy who lived in a hollowed out tree and had a pet falcon. I had always dreamed of being able to really do that. By this time I had been married ten years, and even had children, but my wife gave me her blessing. So about 20 years ago, after having become a minister, I bought two llamas and went on a 1,000-mile llama pack trip. I went into the mountains as a minister, but I came out an environmental activist. I really encountered my soul in the midst of the wilderness.

I hiked up the Pacific Crest Trail, and set the record for long distance llama packing. A bit of a fraud, really, because the llamas carried everything. I actually had a lawn chair, two air mattresses, and an electronic chess set. I carried nothing on my back; the llamas did all the work. I'd see other people with 50 lbs. on their backs, just dripping sweat, while I was sitting in my lawn chair, reading my book, and I would feel guilty.

On my trip, I came over a ridge in Chinook Pass and down into a glade called Big Crow Basin. As I hiked in, hundreds, maybe thousands of crows all flew up and I realized the name reflected the fact that the crows always landed in this meadow. I set up my tent, tied my llamas up and went to bed. In the middle of the night I heard sticks breaking and cries – screams, actually. I was awake. I was scared. I gabbed my 357 and made sure the bullets were all in it. What could this be? A cougar? Was this the sound a llama made when under attack? I got out of the tent with my pistol in hand, to discover a herd of elk. What I had heard was the sound of a bull elk rutting - a surreal cry. I stood in the moonlight and watched this elk and his harem of cows. His antlers filled the space and for a few minutes I stood shivering, maybe in wonder. I was probably naked too, so it's hard to tell, but after a few moments all the elk ran away and I was left awed by this encounter with the wild.

The next day I hiked through the forest and came to the biggest clear-cut I'd ever seen. It took me half a day to find where the trail started on the other side. I saw this contrast, between that wild place and this denuded place. I realized that the elk, whose habitat used to go for hundreds, if not thousands of miles, had been reduced to territory I could walk in two days or fly over in five minutes, and it deeply impacted me. That was a moment I call being born again, again, where I realized that I had also a relationship with creation, with life around me. It was transformative.

"There's a verse, proverbs 31:8, that says speak out for those who cannot speak for themselves. I read this verse sitting on a stump in that clear-cut and thought, "Lord, who speaks for your elk?" The entire elk habitat is gone. "Who speaks for your salmon? Who speaks for your spotted owl?" That was the transformative moment when I decided I would be an environmental activist."

When I came out of the mountains after four months, it was right in the midst of the spotted owl debates. The spotted owl's an elusive bird. Most people have never even seen a spotted owl, but it had been declared that the ancient forests could no longer be logged because of this endangered species. Suddenly these cultural wars had erupted in my community. I saw many of my Christian friends speaking out in ways that I thought were in error. They would say things like, "God loves humans more than God loves owls." But I had encountered God in those forests. I felt closer to God in the midst of creation than I ever had in my own church.

Interestingly, Search Institute conducted a survey and interviewed 30,000 young people. They asked them when they felt most spiritual or closest to God. The number one answer was time out in nature, the second, listening to music. At the bottom of the list was learning the doctrines of their faith. So, in many ways, I think maybe pastors should do more hiking and less preaching.

I realized that I loved a church that was being disrespectful to nature because they had no connection to it. They had no sense of kinship. So I formed my group, which we now call Restoring Eden, as an attempt to speak out. For the next ten years, my work was focused on what I would call classic environmental issues: protecting forests, protecting rivers, protecting species, but over time I began to recognize that there were bigger issues: energy and climate change issues.

A big change happened when Sir Isaac Newton was able to finally, accurately describe the laws of motion of the planets. It was also the same time that factories were first being developed and watches were first invented. The language about God also changed. We went from God the Gardener, God the Shepherd, God the Father, and God the King, to God the Watchmaker. Earth became known as a well-oiled machine. This is in many ways when the evangelical movement emerged. Wesleyan Edwards still talked a bit about earth as the living community, but from then on, increasingly, within the evangelical community, language about earth became very mechanistic.

It's a strange idea, really, that God would create the earth and then step back and admire it as a watch? When you think about it, machines don't have value outside of their function - outside of their ability to do work. Machines don't have intrinsic value. Machines don't sing praises. Machines don't have life. Machines don't have souls. This is important because the bible says we're the only part of creation made in the image of God, but then the church forgets that humans are also made of the substance of earth. These are big ideas.

I don't speak for everyone in the creation care movement, but we've taken a deistic view that God created a machine that has no intrinsic value. Nobody goes to jail for leaving a tractor out in a field to rust. You still can go to jail for abusing a horse. In a way, Christians today are given a difficult dichotomy; Earth is a goddess or a gearbox. I think there's an alternative. It's earth as a garden. Earth as an organic community. The science of ecology is relatively new and it is only when we begin to understand the intersection of all these systems that work together that we can understand the web of life.

Often times, among those of faith, there is skepticism of science, and one of our responses is to marginalize the messenger. Part of this struggle is because we demand proof from science. Science can't give proof; it can only give an indication, or correlation. We do science a disservice to demand that everything has to be accurate and literal. Is there no room for a God who created a world beyond our understanding?

The bigger issue is simply that science has proven that Earth is the only place in the known universe that supports life, and humans are the only species that have the ability to create or to destroy our own lives. There is so much mystery and wonder in all of that. It makes my soul declares there is a God. So we named our organization 'Restoring Eden,' not because we believed we could actually go back and restore Eden, but because I'm an advocate for the only place that we know of that supports life...Planet Earth.

And as an afterthought, we created a bumper sticker that says "God's original plan was to hang out in a garden with a bunch of naked vegetarians", but I always have to tell folks we aren't promoting vegetarianism, we're promoting naked gardening!"

APPALACHIA FOUND: TUG AND JANE
AND THE COST OF COAL
IN HUMAN LIVES...

One of things that Peter made possible was the time we spent with Tug Smith and Jane Branham. I haven't seen much real dirt down poverty in my life. We may have looked poor after my Dad died, but there was never a feeling that we were poor. When I returned to Cleburne, Texas after being gone for forty years and saw the house I spent my teenage years in. It looked like a worn out shamble, but while we lived on those hundred acres, I felt rich with the land, the little lake, the big tree in the pasture, and with life.

Appalachia, too, always sounded romantic. The word itself was poetry, provoking music and mountains, not poverty. So I was surprised to see that the shacks that were built in the 30's for the coal mining families, when mining was booming, run down the hill, and stand only inches apart from the highway and from each other. It's a miracle they are still standing. They have no insulation from the freezing cold winters or the boiling hot summers. They do define poverty, but when I saw Tug's thread bare house, over crowded with pictures and mementoes, it still gave out a romantic reflection.

Tug is a beautiful Elder and a wonderful storyteller. I understood his stories about how his life changed, from feeling rich when he was younger, even though he actually was poor, to feeling incredibly poor when mountain- top removal began. Tug's stories are heartbreaking.

We haven't heard much about mountaintop removal from mainstream media. But over 500 of our mountains have been blown apart, including Black Mountain, the oldest mountain in American and possibly the world, all for 18 inches of coal. This newest way of taking coal out of the Earth has robbed us of our mountain range and robbed employment from the men who have spent their lives as coal miners.

Tug Smith
Appalachia Elder

"We can't move away, we have no place to go, because we are all useless now," Tug began. "We have been replaced by a method called mountain top removal. They have replaced miners with dynamite and instead of removing coal from inside the mountains; they are removing the mountains themselves. Simply blowing up the mountains and poisoning the streams with toxic metals that kill us makes for cheaper coal. I used to hunt squirrels for breakfast, and play in these mountains as a child. We courted up there when we got a little older. The mountains are all being destroyed now. I miss them. It ain't like it used to be."

The kids in the chapel might have been yawning, but where had I been? Where had the media been? Then we found Jane Branham, an activist with Southern Appalachian Mountain Stewards. Standing defiantly, with fists clenched, on the very spot where it happened, Jane shared the story of a three year old named Jeremy Davidson.

Jane Barnham
Appalachian Mountain Stewards
www.sansva.org

"Jeremy died asleep in his crib," Jane cried angrily. "A group of miners hit a boulder that rolled down a mountain and crashed through the wall of Jeremy's double wide trailer. It crushed the life out of him. The coal company was fined $15,000 for being on a site without a permit in the middle of the night. The foreman was so broken up about the accident that he took the half of the double wide that wasn't crushed and turned it into his office. This is the love of money, that is all this is, and it's the responsibility of all God's children to stop it. That is why I am an activist."

I tracked down singer/songwriter Caroline Herring who had written a haunting Black Mountain Lullaby as a tragic love song, in the words of the mother, for little Jeremy. I cried when I heard it. Black Mountain was Mountain Top Removal's oldest North America's victim; Jeremy was one of the youngest. And that was the beginning of the film taking shape as a heartfelt song to humanity.

There were other gut wrenching stories: Sarah Yonts, one of Peter's students, had conducted door to door wellness surveys to determine how pollution had affected the residents of the area.

Sarah Yonts
Student

"In the tiny town of Printer, Tennessee," Sarah began, Located there in the midst of mountain top removal, the drinking water has become so polluted from the toxic chemicals that wash into the stream, that when we went door to door we found six people with brain tumors. In a town that size there shouldn't be any, but we found six, and every person that lives in the town has at least two other types of cancer as well. It's crazy.

I spoke with a mother whose child was born with birth defects from the toxins in the water. The woman was friendly, but begged off from taking my survey because her sickly child wasn't feeling well, and was ready to take his nap. She wanted to stretch out with him until he fell asleep, so she asked me to come back in a day or two when the child was better. When I returned to that heartbroken home, I learned that the child had never woken from that nap. He had died that very day.

These stories stunned me. Why haven't we heard about any of this? Where was mainstream media? Where was the outrage about innocent people dying just so we can live our life with 'cheap' comfort? What is the real cost of coal?

Our last day in Appalachia fell on a Thursday, the town's weekly dance day. We entered the front door of a hardware storefront and were greeted with the sounds of fiddles, guitars, and the music of Appalachia. Men in big hats with taps on their shoes, signs vowing love for God, coalmining and blue grass music. The Appalachia I had longed for, found at last!

Science as Our House of Cards.

Away we flew to the Rocky Mountains. Scientists were gathering, and we were invited. An International Conference was happening a couple of hours outside of Denver. The object of our trip was to interview an Australian scientist rock star named John Cook. John wasn't always a rock star, but like all good rock star stories, when it happened, it practically happened over night. Here is the story in John's own words:

John Cook
Research Fellow, Climate Communication
Global Change Institute at Embassy of Queens
Skeptical Science
www.skepticalscience.com

"Let me just begin by saying that scientific skepticism is healthy," John began. "Scientists should always challenge themselves to improve their understanding.

Our website has spent the last year and a half looking at climate research. Out of many studies, about 4000, we found that over 97% of them endorsed the consensus that humans were causing global warming. When we published our paper, there was immediate intense interest from world media which surprised us. Scientific consensus has been known for over a decade, but revealed in this way, it was suddenly news to everyone. I suspect people didn't realize that there was such strong agreement among the scientific community. Barack Obama tweeted about my research and his twitter has thirty one million followers, so that immediately sparked a whole wave of interest overnight.

I actually started Skeptical Science in 2007 for myself, just as a personal resource. Eventually, this light bulb went on and I thought maybe other people would find this a useful resource as well. Since then, the traffic's been steadily growing and now it gets about half a million visitors per month.

What really struck me when I was doing media interviews with radio and television stations was how surprised people were about the consensus. The general perception among the public was that there was a fifty-fifty debate. And the consensus was not a new finding. There was a survey done in 2009 that found that the higher the scientists' expertise in climate science, the higher the agreement that humans are causing global warming. The studies went back as far as 1990. Scientists actually settled all this twenty years ago.

So I started asking myself about this gap between public perception and reality. **The main cause is a deliberate, focused campaign to cast doubt on the consensus.** As far back as the late 1980's, vested interest groups have been trying to confuse the public. In 1991 the Western Fuels Association, which is a fossil fuel company, spent half a million dollars on a campaign with the goal of re-framing climate change. In 2000, there was a political lobbyist and advisor, Frank Luntz, who wrote a memo to Republicans advising them that if they could make the public doubt the consensus, the public wouldn't support climate policy. This was the strategy that the Fossil Fuel companies and Republican Party have used, not only to attack climate science, but also to prevent climate action. There's been a twenty-year campaign to confuse the public and it's actually not that difficult to do. Whenever a climate scientist is being interviewed, all you have to do is get a skeptic to argue with them. The skeptic doesn't have to be a scientist, the public just sees one side arguing with another and they think there's a fifty-fifty debate. That's how the misperception has persisted. The big question is what is the motivation for attacking science?

It's a complicated question because people are complicated. A dominant theme for this research is that political ideology is the driving force behind climate denial and opposition to climate action. People who are politically conservative believe in free markets. They don't like industry being regulated by government. They see climate action mitigating pollution as a threat to their industry, as a threat to their economy. Rather than try to come up with market-based solutions to climate change, they deny the science or claim that there's not a problem at all.

My interest is really looking at it from a doubting point of view as a social scientist. When you look at all the various arguments against climate science and climate action, there's actually a number of stages: there is the 'It's not happening' stage, the 'Okay, global warming is happening, but it's not caused by humans' stage, the 'Okay, we are causing it, but it's not going to be that bad' stage and finally the 'Okay, we're causing it and it's bad but it's too late to do anything anyway so let's not bother changing' stage.

When I started Skeptical Science in 2007, I had this expectation that understanding stages of denial would move the public debate along. I hoped people would gradually move from arguing about the most basic scientific facts to arguing about the solutions to mitigate climate change. My prediction was wrong and denial of global warming is more prevalent than ever. It's extraordinary that we're still at that first stage of denial.

One popular version is that global warming stopped in 1998 and hasn't warmed in the last sixteen years. What this argument ignores is what global warming fundamentally is. **To set the record straight, since 1998 our planet has been building up heat at a rate of four Hiroshima bomb's going off every second.**

Tragically, oceans are absorbing more than 90% of global warming, which is causing them to expand, which contributes to sea level rise. What also happens is that oceans give up their heat to the atmosphere. They have cycles where they suck up the heat as cold water comes up to the surface, so the temperature jumps up and down from year to year. To understand what's happening to our climate, we need to look at the full body of evidence and not cherry pick tiny pieces of data to try and see the picture that we want to see.

The next stage in climate denial is denying that humans are causing global warming. The beauty of science is you can make predictions and then you can check them with measurements. The way we are causing global warming is by emitting greenhouse gases into the atmosphere that trap heat. The sun shines visible light on to the earth, and visible light can go straight through the atmosphere; it goes through all these greenhouse gases and warms the earth. The earth then radiates all this heat that gets trapped by greenhouse gases allowing less heat to escape out into space.

Scientists can check if that's actually happening by using satellites to measure the heat as it escapes out to space. What they're seeing is this big bite out of the heat at the exact place where carbon dioxide absorbs energy - a specific, distinct fingerprint of human-caused global warming.

The next stage of denial is saying, humans are causing global warming, but it's not that bad; it's not going to cause that much change in temperature. This is known as climate sensitivity. An example is Arctic sea ice. Normally, Arctic sea ice is white. It's reflective which keeps the area cool. When you get warming, the ice starts to melt and the ocean water absorbs heat. This is detrimental to the survival of coral reefs which are like the rain forests. Whole ecosystems survive around these reefs.

It's true that some areas will be, at least for a little while, better off with warmer temperatures, but, again, you can't look at just the small pieces of the puzzle. If you look at the whole picture, the negative impacts far outweigh positives. Plant growth for instance, depends on the right range of temperatures and water as well as CO_2. Some would argue that elevated CO_2 levels due to global warming have a positive impact, but what we get is heat stress for plants. We also get extreme weather that results in either flooding or drought. And so the net impact, when you add up the possible good from CO_2 as plant food, is negative in comparison to the harm done to the environment. Even forests suffer from global warming. Take for example the pine bark beetle in the Rocky Mountains. That beetle no longer gets killed off during the winter, as it has become warmer, so it continues to kill trees. Did you notice all the dead trees as you drove to this conference center?

The last stage of climate denial is arguing that it's too late to do anything about it so we might as well do nothing. It's a false argument because it's looking at the impacts of climate change the wrong way. The question is, "How bad are we going to let it become before we do something?" Then we can try to minimize that impact as much as possible. An ounce of prevention is worth a pound of cure. Every bit of action we do now to mitigate climate change, to reduce our carbon emissions, to replace fossil fuel energy with green or clean energy, means less impact down the road. It means fewer problems that our children and our grandchildren have to deal with. It's never too late to start cutting emissions. We have already committed to some global warming now, but the longer we go without acting, the more intense the impacts will be."

This very thoughtful breakdown of the steps of people's reaction to climate change surprised me, so I shared with John that Katherine Ross' breakdown of the steps that people go through when they are facing death and dying are exactly the same. Those steps also end in acceptance. "Wow," John laughed, "does that mean the human race is smarter about death than about climate change?" We all laughed again.

I had expected a rock star scientist to speak another language – perhaps spouting a lot of facts and figures that would be hard for a layperson to understand. I had the expectation of arrogance, snugness. Instead John had explained Climate Change in such a way that even a child, a complete newcomer to the subject, could understand. At the end of his interview, he insisted that the real rock star scientist at the conference was Richard Alley, and that we simply had to interview him before we left. As part of his snow and ice studies, Dr. Alley is investigating the history of climate and what it might mean for the future. Here is a snippet of his theory of how we affect climate change:

Dr. Richard Alley
Evan Pugh Professor of Geosciences
Pennsylvania State. University
"We see that the world is changing," Dr. Alley began, "Science shows the world

is changing. If you want to ask, "Is it getting warmer?" The thermometers from NASA show that it's getting warmer, so do NOAH, so do the British, so do the Japanese, so do the Berkeley group with the private funding, in part. If you take thermometers just in the country, they show it's getting warmer. If you take thermometers in holes in the ground, they show it's getting warmer. If you take thermometers taken up on balloons they show it's warmer. Thermometers looking down from space, they show it's warmer. But if you live in the world, my tomato patch is in a different plant zone than when I moved to Pennsylvania because the climate is changed enough that what we grow where is also changing.

What we find when we look at climate change is that the people benefiting from burning fossil fuels are not hurting very much right now. It mostly hurts poor people in hot places or people who haven't been born yet. The people hurting from climate change are not doing much of the change because it certainly benefits people like me, but if you are a poor person in a hot place, or if you're trying to live the way your ancestors lived, things are not good.

We know there are still places on earth that are so cold that the snow is not yet melting. I've been to the South Pole. It's minus fifty. If you went from minus fifty to minus forty-nine, it didn't melt much yet, okay? But if you go look around the edges of the snow and ice, we see melting almost everywhere else. Seasonally frozen ground, permanently frozen ground, the permafrost, and the tundra, are thawing in all those places. The Arctic sea ice is shrinking. There is a little growth in the Antarctic, but it doesn't offset the shrinkage in the Arctic. And yes, in the warmer places you see melting happening; this is especially true for mountain glaciers where we see a lot of melting of the snow packs in the spring. The snow is being very strongly reduced.

People have relied on either those glaciers or that snow pack, essentially as their reservoirs, as their dams. We build reservoirs and save water and then we let it out in the places where we want to use it. A lot of the world however, rather than building dams on their rivers, haven't needed dams because the snow formed dams. Huge snow packs sit up there on the snowcapped mountains way into the summer. And then it melts and puts water in the rivers. The rivers flow all summer because that snow is up there melting.

When we lose snow, we lose the springtime and summer stream run off. That means, if you're a fish counting on water in the river, you're not happy. If you're a salmon coming in from the sea, and you have to swim upstream, the flow is lower. Can you get where you're going? Maybe or maybe not. If you're a farmer and you want to irrigate your field with the flow that's coming down from the melting snow pack, what do you do? If you're in a city, and you want to take a shower or you want something to drink, what you do?

So there are concerns in parts of the world now, because, as we lose the snow and ice, we're essentially losing our reservoirs. Either we have to go and build reservoirs, which costs money and may displace people and other things, or we have to figure something else out. And so, the warming is not yet the end of the world, but it really does push people. It especially pushes poor people who are going to have troubles paying for the dams to make the reservoirs that have to do the job that the snow used to do for free.

What do most of us do when we run out of something? We use business smarts. We invent. We do free market. Challenges are opportunities. We can say, "We can take what is given to us by the world, and we can use it in a way that actually helps everybody around here." We know Texas ranchers, for instance, that are reaching up in the air and pulling money out of it, with wind turbines. The power is there if you know how to use it. The wind and the sun are there and they will power everyone.

We have been given resources. We're not going to quit using fossil fuels instantly but, if we learn while we burn, we can get to an economy that powers everybody in a climate that we can live in. It's going to be new corporations, new jobs, and new inventions. It's going to be all kinds of flabbergasting wonderful things because the opportunity here is to not just power us, but to power everybody.

There's a golden rule 'Do unto others as you would have them do unto you'. This is not just an economic issue, this is an issue of fairness and that's an important thing in my world."

Making this film was changing my life. First it had destroyed my romantic outdated notions of Appalachia and cheap coal. Now it had destroyed my idea of scientists in ivory towers speaking an intellectual language that no one could understand. These were two of the top scientists in the world, and both had spoken very plain English from a deep heartfelt concern for their families, and for the rest of us.

Now, as I put this book together, I have to wonder, how many of you have had an erroneous idea, a suspicious idea of 'science and scientists'.

John Cook had even joked saying that the reason that 97% of them agreeing on climate change is so powerful is that scientists can never agree on anything; that a group of scientists could stand on a street corner for an hour trying to agree on where to eat lunch. I had never imagined that a scientist would have such a loving sense of humor about himself.

One Last Thing…

We had returned, tired but very happy, to the hotel that night, our interviews done. The trip had proven a great success. I plodded into the elevator, alone, and as the elevator headed upward, I heard a very clear voice inside my head.

"Eric hung himself."

An electric shock wave went through my body. I have a son named Eric, and, as fate would have it, he wasn't reachable at that minute. I very nearly cried from shock and weariness, and then did the only thing I could do, tried to put it out of my head and get some sleep.

The next day, before leaving Denver, we were going to visit my producer's friend before going to the airport. We arrived in a Denver suburb at just the right time for all of us to have breakfast together at a small French restaurant. Leaving the restaurant, I commented on the fact that it was some of the best food I had ever enjoyed for breakfast, perfect coffee, perfect eggs, and exquisite potatoes. When I asked how a small French café in a strip mall had gotten so great, the reply was simple, "The chef is also the restaurant owner. He adapted real French food to the American palate." "Wow" I said, "We have a fabulous place in Santa Fe, where a Frenchman has done the same thing." My cameraman, Matt Shultze, was listening and added, "We had. We had a fabulous place in Santa Fe. I don't know if it's going to stay open now that Eric is gone. Eric hung himself a few weeks ago."

Trembling, I was very quiet on the ride back to the house. First of all, I knew and really loved Eric, a beautiful young Frenchman with piercing blue eyes and a ready laugh.

He had sponsored a book signing for me when my first book, 'Gifts of Gratitude', was published.

Walking back into the house I noticed an open Bible that lay on the table. Picking it up and glancing at random, I read this sentence:

> *'Listen to the Voice of the Lord.'*

In that moment, my creative path that follows that 'still small voice' inside of me exploded in wonder. I was certainly hearing that Voice. I just hadn't called it by the old formal name of Lord for a very long time.

Listen Galaxea

Alvaro Cardona-Hine
Author, Composer, Painter
www.cardonahinegallery.com

Galaxea
be kind to Galileo Galilei
who is thrust upon you
in the extremity of his need
grant him permission to poke around
he imagines he investigates
and I love him dearly dearly
he is almost blind
but who needs eyes
when they have you Galaxea
in whose embrace
webs of eternity
rock gently to and fro

take this man Galaxea
to be your lawful wedded mite
and see that you nurture him
and provide for his sustenance
guarding him from the violence
of your real lovers the gods
in whose arms you burst
like a bubble of joy
ah Galaxea Galaxea
to think that it was your own little whim
to invent him this man
what was it?
that you couldn't look with your own eyes
at your armpit?
you couldn't see your own beautiful back
graceful as a beach
and you wanted someone to walk
its lovely loneliness a while?
a while only?
listen Galaxia
let me
let him caress you
let our lives fall upon your skin
like a prolongued caress
this is man
with a day's beard behind him talking
let him care for you
while someone awakens

the sleeping horses
Galaxea
I have used one of your days
to comb the beach
for the humen of doors
it is dark now
I know that Galileo will stumble across
marvelous things in that same dark
which is the mild of Glaxia
Glaxia
if it is within your reach
protect him now that you've made him
pollen man
dandelion man is among you
he is little people
in the eyelash of oblivion
oh Glaxia
see to it that he lives
that he doesn't lapse into chair
that he doesn't sink into stone
that he doesn't faint into thin air
because Glaxia
Glaxia are you listening?
you have no one else smaller than you
to love you
and there is no one else
at the dark beach of your back
running like the wild surf with his lungs
and his now awakened horses

NEXT STOP, TEXAS TECH

By the time Aaron sent us to our next interview, I was already on shaky and forever changing ground. We were meeting with Katherine Hayhoe, the national poster girl for Climate Change. Here came that still small Voice again ... because in order to get to her, my cameraman, Matt Schulze and I had to drive through Texas.

"You can cover up rose, you can cover up Yalla,
but don't you cover up Texas!"

Stan Freberg,
The Yellow Rose of Texas, 1955

What had started out as just a job was taking me through layers of memories - both sadness and joy - that I didn't realize were waiting to be re-discovered inside of me.

I was 'born and raised' in Texas. I was that kid that had insisted on sitting at the back of the bus when it wasn't cool. I was once expelled from school, and even had fist fights over racism. I was that girl who hid in the barn to escape sexual abuse. I was a total misfit in Texas, and had fled the state as early as I could, and had not chosen to return…until today

We arrived late in the evening and found one of the only places still open for dinner. As we were being seated, I noticed that the faucet on the side of their open kitchen was gushing. Not dripping mind you, but gushing. What surprised me was my sensitivity to it. I immediately stepped forward to speak to the hostess, and the visible faces watching from the open kitchen, even as our waitress snapped at me, "Lady, the main kitchen closes soon, I need to get your order turned in." "Give me a second," I explained, "that faucet is gushing water." As everyone started laughing, I quickly realized that the joke was on me. A faucet gushing water, so what? It was only water.

Welcome home. I felt thirteen again, wondering why no one else seemed to care. I expressed my shock while Matt ordered his dinner. If he had the duct tape from his camera kit, I am sure he would have used it on my mouth. Fortunately for me, the equipment bag was in the car.

The Climate Change Poster Girl

Katharine Hayhoe is suddenly appearing everywhere. She is almost always included in TV and magazine interviews on Climate Change, and has even been featured in Showtime's program, 'Years of Living Dangerously.' She is a woman of faith, and the wife of Evangelical preacher, Andrew Farley. With all of that, you might expect an older woman with her hair in a bun. Instead, she is a young, attractive, vital woman who is obviously a mother and a homemaker as well. Take a look at these credits.

Katharine Hayhoe

Author, Climate for Change: Global Warming for Faith-Based Decisions
Nobel Peace Prize Intergovernmental Panel on Climate Change
Associate Professor, and Director of the Climate Science Center.
Texas Tech University

"As Mark Twain once said, climate is what you expect and weather is what you get," Katharine began laughing, then she added a confession, "Being a climate scientist is a hazardous profession these days. Climate change has become very political. Many individuals want to think it is not a real problem. We scientists think of ourselves as being in the trenches; if you stick your head up over the trench you tend to get it blown off. When you go to scientific conferences, it's not just questions about science, it's questions about how to deal with litigation, how to deal with harassment, even how to deal with the death threats. Because people have decided it's easier to shoot the messenger than listen to what we have to say.

But it's important to understand what it's all about, so lets begin by understanding what climate change really is. It's not that one super-hot day, or that freak ice storm, it's not that unexpected tornado, or even that dust storm. Climate change is the average statistics of weather, long term, over 30, 40, or 50 years.

Our civilization, our entire society, is built on an assumption of a relatively stable climate. The places we grow our food, the way we build our houses, even where our biggest cities are located right on the edge of the ocean. All of that assumes that we may have cold years and we may have dry years and we may have wet years, but over climate time of 20, 30, 40 years, it all averages out. That's the assumption our society is built on.

Today that assumption is no longer valid because our average climate is changing. And it's not just about warmer temperatures; it's about rising sea levels as well. Two-thirds of the world's largest cities lie within a few feet of that sea level. It's about changes in our rainfall patterns, where some places that need water are getting drier, while other places that don't need any more water are getting wetter.

If we build our houses for a certain type of temperature, and it gets a lot warmer, our energy bills go through the roof. If we build our cities assuming the ocean won't rise and flood our cities, what happens if it does? We're simply not prepared for these types of changes.

Today we know that our climate is changing because of thousands of different ways that God's creation is telling us that something is different. Our air temperatures are getting warmer around the world. But we also know that birds are starting to migrate at different times. We see bugs in our back yards that we didn't see until you drove 200 miles south of where you lived. We see changes in snowfall patterns, in rainfall patterns. If we look all around the world, there are 26,500 indicators of a warming world. God's creation is telling us that something is different today. Something is unusual. Something is changing.

Whenever we see climate changing, the first thing we do is we look to the usual suspects to see if they can be blamed for our current trends. We know that in the past climate changed because we got more or less energy from the sun and that made us warmer or cooler. Climate also changes because of natural cycles, and over time the shape of the earth's orbit around the sun can change. It can become more circular or oval shaped which affects how much energy we get from the sun, and puts us in ice ages or warm periods like we have today. We also have natural cycles inside our earth's system like 'El Niño' that rearrange our heat between the ocean and the atmosphere, sometimes taking heat out of the ocean and putting it into the atmosphere, other times doing the opposite.

So when we see our air temperatures getting warmer we have to wonder if it could be a natural cycle. Could it be the sun? Well, in reality, the sun's energy has been decreasing over the last forty years, so if the sun was actually controlling our temperatures we would actually be getting cooler, and we are not. The sun has a perfect alibi.

What about the changes in the earth's orbit? Are we still getting warmer since the last ice age? Well no, if we do those calculations, it tells us that over the next thousand years, if we're here that long, we can expect another ice age.

What about natural changes within the earth's system? According to the 'El Nino' phenomenon, if the atmosphere is getting warmer the heat is coming from the ocean or vice versa. Today, both the ocean and atmosphere are getting warmer. It can't be a natural cycle. They all have a perfect alibi.

So if the usual suspects can't be blamed for our current warming, what can? Scientists discovered, almost 200 years ago, that the earth has a perfectly designed blanket of invisible heat-trapping gases. This blanket keeps the earth almost 60 degrees Fahrenheit warmer. Like a blanket traps our own body heat on a cold night, this heat trapping gas blanket keeps the earth at a perfect temperature. Scientists also discovered that whenever we burn coal, gas or oil, we release a lot of carbon dioxide into the atmosphere that would otherwise stay trapped in the ground. Carbon dioxide is one of the main gases that make up that invisible blanket. Essentially what we're doing is wrapping an extra blanket around the earth, a blanket that is not natural.

Many of us believe that God has expressed Himself through His written word, the Bible, as well as His created world, the Earth. In studying the science of God's creation, I believe that I am studying what God was thinking as He set up this amazing plan. But Christians have been very wary of science because there have been many conflicts, some going back hundreds of years, between science and faith. People point to the example of Galileo and the Catholic Church, of Charles Darwin and Creationism. Or even recent issues such as the sanctity of life; when does life begin? What type of stem cell research is appropriate? On many issues like this, people of faith and people of science have often fallen on different sides of the issue. So along comes climate change and people say, "Oh, well it's just another one of those Godless, liberal, atheist, tree-hugging issues, clearly not anything any self-respecting Christian would ever believe in. But the issues we have with climate change is based on scientific facts and data, and our calculations now tell us that climate is changing, and that humans are causing the change.

Science can tell us how the changes are going affect our health, our water resources, our food, and our wallets, but that's the point at which science stops. Because what to do about climate change is about choice, it has to do with what's in our hearts and what our values are. And for us as Christians, the biggest value we have is the value of love. We're told to love our God and to love our neighbors as ourselves. So here we have a situation where we are contributing to an enormous problem that is harming people - the poor, the needy, the vulnerable – the very people we are specifically asked to care for, who will not be able to adapt when climate change affects their livelihood. So I say it's an issue of love to care about climate change.

I believe that the real problems we have with climate change are the solutions. We hear about increased restrictions, government legislation, and taxes. None of these are words that make us feel very happy, let alone excited about taking action. More importantly, even though science can tell us what the outcome of a given solution would be, it can't tell us which solution to pick because again that's a value choice.

There is no silver bullet that will fix the entire problem. Individually, we don't have the incentive to reduce our own contribution to the problem because many of the resources to do so aren't available to us. I would love to paint the outside of my house with the solar paint they invented a couple of years ago. Who wouldn't like to do that? Free energy. I would love to put solar shingles on my roof. I would love to have an electricity bill of zero, especially in the summer in Texas. But I don't have the resources to buy these things because they aren't widely available yet.

So I put a lot of effort into trying to reduce my own contribution to climate change through reducing the amount of energy I use. Replacing light bulbs, buying more efficient appliances, and buying a smaller car that uses less gas are ways that help, unfortunately I am not able to see the immediate benefit to the environment. Only if everyone acts together will it be a big enough change to witness with our eyes.

Many of us feel the solutions to climate change run against our values. We feel as if the solutions involve a loss of personal liberties, economic hardship, restrictions, and higher taxes. What we have to realize, though, is that there are solutions to climate change that bring us personal benefits such as growing our own energy here at home instead of depending on foreign countries. What we get are energy sources that are renewable, that are not going to run out on us. The one thing we know is that coal and gas and oil are finite and they will run out one day. A renewable source of energy builds our local economy. There are a lot of things I believe we can do to make our lives better, things that improve the quality of our communities, and incidentally, help with climate change.

With the issue of climate change there's often some theological questions that come up too. People frequently ask, "If humans could change something as big as the earth, aren't you implying that God's not in control anymore?" Not at all. Of course God is in control! But God has given us an enormous amount of power through free will. We can make good or bad choices, but we reap the consequences of them. There are consequences for decisions, but they are not the same as God's punishment.

The first choice we can make is to increase our efficiency. That simply means let's stop wasting so much energy. The United States is one of the most wasteful countries in the world in terms of energy. We leave the lights on in entire office buildings all night. If we would just turn off our computers at night when nobody's using them, we would save enough electricity to power several small states.

The other choice we can make to solve climate change is to transition away from the dirty, and frankly, outdated ways of getting energy that we use today. Coal has been used for over a thousand years. In the year 1200, so many people in the city of London, England were burning coal, that the king enforced the first air pollution laws in history. That was over eight centuries ago, and we're still getting our energy that way today. If you were given the choice between a really cool shiny new iPhone and one of those old clunkers that literally weigh 20 lbs. which would you prefer? Of course everyone would want the shiny new one. Well, the shiny new way of getting energy is renewable energy - energy from the wind, the sun, tides, and geo-thermal energy from the earth.

Currently, our entire society is built on the old foundation of coal and gas and oil. Transitioning the foundation of our society is an enormous effort. Yes! Things will be better, but man, it takes a lot of work and a lot of effort, and a lot of investment to get there. A good analogy is to look back in time. Look at when society in the

South was built on the foundation of slave labor. People could not imagine how the South could be prosperous without that base. And so some of the very same arguments to retain slavery that were made back in the 1800's are being made today to retain fossil fuels. It's too expensive. It's not that bad. Nobody really minds. These are literally the things that people were saying back then to preserve slavery and people are saying the same thing today to preserve fossil fuel. Today if you calculate how much energy each of us uses individually, it's actually the equivalent of about 200 slaves or more. Imagine replacing that amount of energy with a new source of energy from the sun or the wind or the tides. It requires a shift in the basis of our society that is as big as or bigger than the shift that took place when we abolished slavery because we realized it was wrong.

Climate change is a daunting problem and I think part of the reason why we're in the situation we're in today is because, if we really open our eyes and we look at the scale of the problem, it's scary. And what does fear do? Fear paralyzes us just when we can't afford to be paralyzed. All around the world we see examples of how climate change is already affecting real people's lives today. In the Arctic, in places like Kivalina, Alaska, Inuit villages are built on what used to be frozen ground. Today that ground is melting and crumbling into the ocean, house after house disappearing into the sea, hundreds of years of traditions failing them. The ice no longer supports their weight. The animals no longer come when they used to. The frozen lockers underground that they used to store food are melting and the meat is decaying. If we can see it happening to real people, we can grasp that it is not a distant issue that affects the polar bear, or only our grandchildren or great grandchildren. Climate change presents us with an opportunity to care about others now.

The issue of climate change has become extremely politicized in recent years. But if we believe that whatever we reap, we sew; if we believe there are consequences, not punishment, but consequences to our choices, Climate Change is not incompatible with faith. It may not be compatible with our politics but it is completely compatible with our faith."

Wow, this is the woman guiding the Nobel Prize environment committee? I have been too embarrassed to even say the word God for most of my life, and she, as a world-renowned scientist, is saying that our duty to God is to care for the poor whom we are hurting with our lavish lifestyles. That what we choose to do about climate change is a question of love.

In order to even get Katherine to do this beautiful interview, I had to sign an agreement that stated I wouldn't use her interview if I even suggested that climate change was still a 50/50 debate. I had begun to understand the scientist's frustration, 97% of all their studies prove that it's real, and it's coming at us! And I didn't know if my samplings were representative, but so far, every single person who was crying out about the dangers of climate change was a sweet, kind, intelligent person who was genuinely worried about the human race, about strangers that they didn't even know in other parts of the world. What a surprise!

It was about this time that my cameraman, Matt, began to ask me how I was going to put the film together. For once, I was a complete blank. In documentary film making it is customary to show both sides of the argument, but this sure wasn't going to be documentary business as usual.

On our way out of Texas, Matt selected to stay in the car when I stopped again, at the IHOP, to give them hell about that faucet.

We Head For Washington D.C.

I had been to D.C. once before. My memories were of the magical trip I made when a man of color got elected to The White House. For the little girl who had heard "not in your lifetime" a million times about people of color, it was a trip not to be missed. Now here I was again, and this time people were actually scoffing. "Nobody is interested in climate change," was the discouraging announcement that I was hearing from most.

I shudder to think of what might have resulted to my soul had I stopped to listen. At the time of all this discouragement, there was no way of realizing that Climate Change was about to boil over into being the hottest contested topic of our time, and that by the time the film would be finished, 2015 would be the hottest year in recorded history.

We had come to meet another rock star, this time an Evangelical. Time Magazine named Rev. Richard Cizik one of the top one hundred most influential people in 2008. He is a man on fire, and is almost always in the middle of one sentence or another.

Reverend Richard Cizik
President: New Evangelical Partnership for the Common Good
www.EvangelicalPartnership.org
Lobbyist, Washington D.C.

"I think it's possible to get there, with new understanding," Richard began, "but there has to be some type of conversion. That is what happened to me in 2002 at the Oxford Conference on Climate Change. I was college educated, masters' educated. I had an honorary doctorate in Christian leadership. I was top lobbyist and an advocate for Evangelicals climate deniers in Washington. In other words, I was at the top of my game. I was at a conference when my mind suddenly clicked and I said, "Wow! This is so fantastical. I'm seeing something I've never seen before! How could it be that I didn't understand this?"

I had gone to the Oxford Conference against the advice of some of my elders. Those in the organization said, "Why should you go?" and I said, "I've got to go and hear the case to be made for global warming." I went to Oxford and I had something that could only be described as an extraordinary conversion - that is a total shift of thinking - because, for the first time in my life, I began to see the world differently and it was so fantastical that I could only describe it as a conversion.

And my friends said, "Well, Richard, you're a born again Christian. Can you describe this shift as the equivalent of your saving grace and salvation story?" And I said, "I most certainly can." This is the most important shift I would suggest anyone can make post - conversion. That is, to see the world as Jesus sees it requires us to have a new sight, not just to see things, but to behold them differently. What we have to do, as Evangelical Christians, is see the world completely differently, see it through God's eyes. And we're not doing that yet.

For me, the conversion happened because I suddenly beheld the science. That's not the same as the theology. The theology came next, but it was actually the science of climate change, the evidence of what's happening to the planet, that is irreversible, that so struck me that I was hit in the face, where I had one of these "OMG" moments, and I said, "This can't be really true. If it is true that the planet is changing and simply can't be fixed by cleaning up a river or taking the pollution out of the air, and that through CO_2 emissions being released into the planetary ecosystem, we are creating a greening effect that is warming the planet, that is changing it in ways that will forever be changed, if that is true – really true - then we have to change everything. Most importantly, as Evangelical leaders, in positions of teachers and educators and pastors, we have to teach our flock the same things lest we fail God in the most important of all capacities, that of people who understand the times and teach accordingly."

For people to make that fundamental shift in their worldview, to behold it differently, they have to be confronted. They have to be challenged. There have to be disturbers - people who will make somebody rethink the way they thought the world was - so that they change. Eighty percent don't want to do it, and they won't shift their worldview unless there's somebody who says in the midst of their status quo "if you choose to survive, you're going to have to change." That role of the disturber is fundamentally the role, of the preacher and the teacher.

Making the shift to sustainable thinking, to seeing the world differently and living according, is not something just for some. It's for everyone. Because we've gotten on this roller coaster and allowed ourselves to go very, very high, living in an unsustainable way. We're at the height of the roller coaster and we have no idea at this point how fast or how dangerous our fall is going to be. But it could be very dangerous, and catastrophe may occur, because we've been living in ways that defy God. We have been defying how He said to live.

It is the eyes of our hearts that see differently. If we can't see the world differently than we're not going to behold it and take in these truths that God wants us to have. If you're rejecting climate science simply because scientists are saying it, scientists who are largely evolutionists, if we allow that kind of illogical syllogism to determine how we react to the science, then we are really foolish people. Because we need to listen and learn from scientists about what they're saying about creation, because creation can't always speak to us. But scientists can tell us what creation is saying, and creation is speaking out, it's crying out to us in so many ways. It's just we don't have ears to hear. I was literally like John Newton, the slave ship owner, who said, "I was blind but now I see."

After my conversion, I spoke out on these issues and annoyed existing religious right leaders because they couldn't stand to hear what I was saying. But I wouldn't shut up and eventually I lost that job. I am an Evangelical. I remain an Evangelical. But what has shifted is the way I view the planet. I began to see how bigoted and narrow minded, how hardnosed and dug in we as Evangelicals are and have been, and very importantly, how other people around the planet view us. It's really important to get outside of the box and go to other lands and see how they look at us, in our malls, in our affluence, living a life of affluenza, as it's sometimes called, the kind of flu-like sickness, while the rest of the world suffers. We can't drive our SUV's oblivious to how CO2 is impacting somebody in Saharan Africa, India or in the Himalayas

People ask me, "Well, where do I begin this shift?" and I tell them that it sort of depends on where they are in life. A college kid will say to me, "I don't have the resources to buy a hybrid or whatever." And I say, "Well, you don't have to - you can shift something today that will impact the rest of the planet. You just don't know it." And they ask, "Well what is that?" And I say, "It's what you eat!" That's one thing every one of us can do because to live the lifestyle we in the West live, especially with our consumption of red meat, we're putting demands on planetary resources in ways that are unnecessary. So it begins with something as simple as what you eat. Of course all the rest is important too. You can shift your transportation. You can shift the way you use resources. It all begins with seeing the world differently and making little changes that become big changes. It's from I won't, to I might, to I will, I am, and I have done.

That shift is sometimes gradual, and sometimes it's very quick, almost cataclysmic, like a lightning bolt that illuminates your mind and you immediately begin to change. But regardless, you're not off the hook just to say, "OK I know the world and the planet's resources are suffering because of the way I live." Once you understand that you have to take action.

So now I'm part of a movement called New Evangelicals, and we're simply trying to reclaim what it truly means to be born again and be a follower of Jesus. Which, in a nutshell, means you love. You don't hate. We believe all the issues matter, not simply issues like abortion and same sex marriage, which is what the Religious Right thinks. No; human rights, economic justice, the environment, and protection of the family - all of these issues matter. So, it's a shift from thinking that I live in order to die and go to heaven to be with Jesus. No, no, no! I live in order to be a co-partner with God in the redemption and renewal of all of creation. That is the shift. That is the real calling. That is why we were put here and yet for two thousand years plus, most 'Christians' have gotten it wrong.

An evangelical theologian, such as Anne T. Wright, whose new book on heaven says exactly what I've said. It explains that we've followed Plato; we've not been following Jesus. And what did Plato say? He said matter doesn't matter, and we've lived as such. The earth doesn't matter. It's all going to be burned up. As a preacher in Seattle said recently, "Drive your SUV because it's all going to burn up." Nothing could be further from the truth. We are called by God to be co-partners with Him in the renewal of all that He created. That's a different story than dying to go and live on a cloud.

Your responsibility from Genesis to Revelations is to care for what was given us. Because when you die that will be the question....."What did you do with what I created? I gave you gifts and abilities and what did you do with them?" Did you preserve what I created, or did you just trash it on the basis that it is going to burn up?" By the way, that is the worst interpretation of the scriptures that I can think of, burn up? No, no no! That's not what the scripture says. It says this world will be refined in the end. Refined but not burned up.

So we have to shift from, we take things from the earth, we make things and we put the trash back into the earth, which is the take, make, waste mentality, to a borrow, use and replenish mentality. This says we borrow from the earth, we use its resources for God's glory and then we replenish the earth. We can live that new way, and it's entirely plausible; people do it. That's our responsibility. That is exactly what the Bible says we are supposed to do all of our earthly lives.

Supposing a man leaves on a boat to go from point A to point B, and on this boat there are people who are equipped and prepared to feed others, to care for them, their health, minister to them in every way possible, but, out in the ocean, the captain is confronted by one of his staff who says, "Sir, we have a major problem. The boat is taking on water and we are going to disappear in this ocean, never to be seen again, unless we fix this." That shift on the ocean is happening to us, and the boat is called planet earth. If we don't take care of planet earth than all the other ministries we're engaged in, they don't really amount to a hill of beans, as the expression goes. Because, if the ship goes down, if the planet is destroyed, all those other ministries we are engaged in are simply trying to take care of the people on the Titanic without caring that the ship is steaming toward an iceberg. And, once you hit that iceberg, you have to ask yourself, "How could we not understand that our fundamental responsibility was to care for creation?"

I was approached by the Environmental Working Group to have a test done of my body called a "body burden" test. I said yes, I'd like to know what composes me, and what toxins I have in my body. My blood and urine were shipped to five different testing labs around the globe and the report came back in the form of an actual book that tells me exactly what's in my body. Of the over three hundred toxins and chemicals that exist, I have over two hundred and eighty of them. I haven't ever worked in a chemical factory. I'm a pretty average American in that sense. I haven't been exposed on a daily basis to these kinds of chemicals and toxins and yet, the truth is, I have plasticizes that come in the furniture we're sitting on, and methyl mercury that comes from the coal burning utility plants that emit this on a daily basis through the air. Every one of us is probably, to some degree or another, just like me. Filled with all kinds of chemicals and toxins. And then we wonder why we have the cancer rates, cardiovascular disease and infertility, which my wife and

I experienced. Infertility, I came to find out, comes from mercury, among other things. So the cancers that we have, the diseases that we confront on a regular basis here in America, they come in part from all of these toxins and chemicals that are emitted into the world around us, and you know what? We do it to ourselves.

We permit it through the political process, wherein multi-million dollar corporations buy politicians. It's just legalized bribery. I'm a lobbyist, and have been for thirty years in Washington, and I can tell you that corporate enterprises buy the Congress of the United States. They get away with not putting warning labels, they get away with emitting mercury and methyl mercury through coal burning utility plants, and we act oblivious, but there are consequences.

Fortunately, the Environmental Protection Agency, under the authority of President Obama, pushed through the Clean Air Act, and has been forcing the industries to make these shifts, to put scrubbers on their utility plants, to reduce mercury and methyl mercury emissions, which by the way, produce impacts on the unborn. One out of eight children in America are born with impacts, including mental retardation that comes from mercury and methyl mercury, but businesses are only doing it now that they're being forced to do it by our government.

Let's be honest, the Republican Party is always fighting against these restrictions because it's costly to their friends and supporters - the utility industries that pay their campaign bills - and so they're always fighting against the EPA. It's all about money, shareholder money, campaign dollars, and corporate profits. That's what matters most, and it's a crime. The structural sin in America is one that puts a priority on profits at the expense of any other human values and one that is contributing to the climate change that is so real, with its impacts so great on the rest of the planet. When we are simply 4% of the planet and contributing to over 25% of the world's greenhouse gas emissions; that is called structural sin. We are behaving like foolish people who think we can live pious lives in our little narrow worlds in America, with all our affluence, and our SUV's, and our home air conditioning, while the rest of the world in places like Delhi are at plus hundred degrees Fahrenheit temperatures. I was just there and it's hot. And the planet is getting hotter.

It's the worshiping of a golden calf called our lifestyle, the lifestyle that has been given us and we think we're owed. God will hold us accountable. That's why Revelation 11:18 says, "I will destroy those who destroy the earth." It's not somebody else who's doing it. It's what we are doing to ourselves.

We behave as if these generous resources that have been given us in America are infinite, and they're not. They're finite and they will run out, because climate change is going to blow our minds. We are on this roller coaster of our own making. We're at the top of the ride and we don't know how far and fast or when it's going to drop, but it will, because you can't bribe Mother Nature. You can't cheat Mother Nature.

We're seven billion people now, and, at current rates, we're going to soon be ten billion. Can we even sustain a planet with ten billion people? So I say, why shouldn't Evangelicals support choice? It should be a mother's choice as to whether or not to have a child. We should support family planning. Family planning doesn't equal abortion. Family planning equals the access to contraception that makes it possible not to have a child, or at least, to have it when you want to have it. The fact of the matter is, around the planet, one woman dies in childbirth every minute.

That's five hundred thousand women a year that die in childbirth because she doesn't have the simple access to what we take so for granted in the United States, which is family planning or contraceptives. So we have to pay attention to what's happening in maternal and childbirth around the planet and we should change our views about family planning. Instead of viewing it as abortion, which it is not, we must view it as pro-life. It prevents abortions. It prevents maternal deaths. It saves children. Family planning is an Evangelical value in my estimation.

We have to change and, moreover, we can change. It's all eminently doable if we will open our hearts, our minds and see what is happening. I know we can change because I've changed, and if I can change, as stubborn and as unworthy as I am, believe me, anybody can change. I was as conservative as could be and as inclined to think I knew it all, and then I had a conversion to the science of climate change that then led me to change myself.

The earth is resilient. It changes. There are extinctions that occur all the time. But that doesn't mean that we humans should be responsible for a lifestyle that contributes to extinction. Is the planet able to survive one species that then begins to dominate and destroy all the others with climate change, habitat destruction, and pollution, the destroying of all of the resources that were given us by God? Is it right for one species, us, to do that to everything else?

So I'm for divestment. That's my answer. I go around the country asking people to take their investments out of big oil because that's the only way the oil companies will get the message. If you believe in caring for your neighbor, then what better way to care for your neighbor in Saharan Africa that's already suffering from climate change, then by taking your investments out of those companies that are responsible?

Divestment is responsible. It's biblical. It's compassionate. It cares for the planet and it should be done. And believe me, the CEO's who seemingly thumb their noses at us as if we're stupid will get the message. Because we're not stupid, we do get it, and we can stop it."

When the fiery, deeply personal interview with Richard Cizik was over, I realized again that my life had been changed forever. The ecstatic love that I had felt for 'God" and the Jesus that I had known and who had guided me as a child, had been reborn, restored. 'Wow!' I thought,' I've been a "New Evangelical," for a long, long time.'

To Stop the War, Make it Green

I admit it. I hadn't done all of my homework on Brigadier General Steven Anderson, but I did know he had an enormous amount of information about the wars we fight. I was still reading his lengthy bio, when a quiet man came into the empty room and took a seat. I was astonished...responsible for 130 US/NATO logisticians in Iraq and Afganistan with a $5.5 billion annual budget. I didn't want to waste one word on 'small talk.' I immediately wanted to know how he did it.

Steven M. Anderson,
Brigadier General, retired, United States Army
Chief Executive Officer
RELYANT, LLC
www.gorelyant.com

"I became an energy warrior because General David Petraeus made me one," Steven began, "My requirements, or responsibilities, when working for him, was to provide all the logistic dependency for our soldiers in harm's way. At the time over 300,000 soldiers and contractors were serving in Iraq and Afganistan.

One of the things that General Petraeus wanted us to do was get traffic off the road because our trucks were our greatest threat. Most of the deaths at that time, and to the present day, occur on our roads while moving things, people, supplies or equipment. The enemy knows this and they plant bombs – and they kill us. So how do you beat a bomb? You beat it by not being on the road in the first place.

When I did the analysis, and I started looking at the number of trucks on the road, I found that four hundred or so of those were doing nothing but moving fuel - the vast majority of the fuel going to power generators that drove electrical energy. And why did we need so much electrical energy? The answer is that we were essentially air-conditioning all of our structures in a very inefficient way. So I became an energy warrior to reduce our dependency on our oil pipeline in order to get trucks off the road and save lives.

Since that time, I've become increasingly more frustrated and disappointed that the Department of Defense, as an institution, hasn't come to the same kind of realization that I have. They aren't doing the things that really need to be done to save lives and to reduce the burden on our taxpayers. Energy efficiency is a nice thing to do but it's not a requirement, and until it becomes one, we will continue to work around the edges but won't make any true changes to the way in which we consume our energy.

Perhaps the biggest roadblock to all this is what I call the tyranny of passive leadership. You've got people, decision makers in key positions of authority, that are afraid to make a bold decision that puts them at risk or exposes them to criticism. There are some exceptions. Secretary of the Navy, Ray Mavis, understands that the navy needs to change by developing a green fleet, and using bio-fuels, but generally speaking, there are authorities that are paralyzed by fear.

One of the big things the army's been very proud of is the use of electrical vehicles to replace Ford F-150s. They have essentially replaced a lot of big trucks with electric golf carts, which is great, but has a very small impact. The net value of all the electrical vehicles has essentially saved the equivalent of six hours of energy use in Afghanistan. Six hours. So, is it important? Yes. Are there more aggressive things that we need to do? Absolutely. That's the point; we need to do more. I focused on our fighting in Afghanistan and Iraq because that is my experience. That's where we really need to cut our dependencies. We need to develop alternatives to moving fuel to sustain air conditioning.

A rolling fuel truck is a Taliban target. Who couldn't hit that? If we didn't have so many fuel trucks moving on the road, we'd be able to focus more on winning the hearts and minds of people that we're actually over there to try to help. I wrote this explanation in the New York Times and I got all kinds of emails. One of them was from a company commander in Afghanistan who told me that he was running a small installation, a forward Operating Base, in Afghanistan, and he had to completely shut down his combat operations every two weeks for two days to go back to his battalion headquarters to get all of his fuel and bring it back. One day to go down there, another day to come back. During those two days, not only was he exposed to enemy fire the entire way, but upon return he had to start all over again because the enemy had reset themselves in the same positions that he'd just spent two weeks to get them out of. He said it was like Groundhog Day - you know, the movie - the same thing over and over again every day. Why? It's because we have this incredible dependency on oil.

So what can we do as a nation? Well, what if we gave that commander solar capability, wind generation, or some of these micro hydroelectric plants? Are we going to put this company commander in a Prius? Well no, but what if we can do things to reduce his dependencies on electrical energy? What if we use something Afghanistan's got plenty of - solar exposure? Think about the impact, how many more soldiers might come home alive. Think also about having an operational energy test bed in Afghanistan where we learn about renewable energy technologies. We're developing these technologies in an operational sense and now we would be able to exploit this capability and bring it back here to the United States.

In order to start turning the ship in the right direction and make the Department of Defense truly capable of empowering our nation to change, we need a policy, signed by the Secretary of Defense. By making it a requirement, commanders would become responsible, accountable, for their energy use. Right now there's none of that. I'm asked the question a lot, "What can Americans do?" I tell them to use their constitutional powers. Their elected officials have got to make the right choices and get on the right side of this issue. Their voices need to be heard.

You asked earlier, "What's the cost of oil over there?" Let me give you some specifics: the cost of oil and fuel in Afghanistan to taxpayers is twenty billion dollars a year. What if we were able to reduce that requirement by say thirty percent? That's five, or six billion dollars. Think about what this nation could do with five or six billion dollars.

The military is an incredible conduit for change. The military does many things that can help America develop as a nation. Think about what they did, in terms of developing rocket technology, to help us put a man on the moon. We put that man on the moon in 1969 because our entire nation was committed to the effort. That shows the power of this nation when focused. When we get a President and a Department of Defense and a Congress that stands behind a green energy economy, and says, "We're going make this happen." I guarantee that will set the conditions for the natural, entrepreneurial spirit and incredible capacity of this nation to innovate.

What's sad about these wars is that, for the first time ever in our nation's history, we are funding both sides of the war. Because of our addiction to oil, our money goes into the coffers of those who would do us harm, the people building bombs to kill our soldiers. We're sending a billion dollars a day to Iran. That money goes into our enemies' hands.

"Drill, baby, drill" is not an acceptable answer. The problem is not the source of our addiction. The problem is the addiction itself. You don't get a heroin addict off of heroin by providing him cheaper heroin. The answer is to get him off the drug. In our nation, we have a drug - it's called oil. We need to get off oil. We need to solve the addiction. Finding more sources to sustain that addiction is not the answer. The addiction is creating climate change and anybody that doesn't understand climate change is, in my humble opinion, an idiot.

Climate change is real, it's occurring, it's a fact. We need to address it. Getting off of oil is how we're going do that. Developing renewable energy technologies is how we must do that. Climate change creates instability. And instability creates political upheaval and tension. And when that happens, then instability unfortunately, puts soldiers like myself in some foreign place to fight and die in order to maintain the peace."

In a little over an hour, Steve Anderson had performed yet another miracle for me. He had changed forever my idea of what being a soldier is all about.

Peace

Hawah Kasat
artist, author, community organizer
One Common Unity, a non-profit

Hold close in your heart
This simple understanding and dream
Because no matter how far away it may seem
or how lost we may be,
The light will always reach us
and love will always find us.

No matter where we have buried our hearts
Love will find us, Love will find us,
And make miracles of us all.

Contrarians

A true definition of ' contrarian' is people paid to make you doubt. I kept hearing about the people planting doubt in the mind of the public about global warming, and was very curious, not only as to who these people were but why they were working so hard to debunk climate change. I didn't have long to wait for answers. In DC, we met Rachael Tabornick, a woman who has spent years researching the 'contrarians.' Rachael connects the dots.

Rachel Tabachnick
PRA Associate Fellow
Presentations on the impact of the Religious Right
Follow her at @RTabachnick.

"We're going to start in a place that may surprise you," Rachael began, "We're going to start with the business and politics side of the story. Let's look at the tobacco ads in the 1920s. Understanding their impact will allow you to see how it leads directly into the controversy over global warming. Remember: "More Doctors smoke Camels than any other cigarette?" In 1998, as you may recall, after decades of continuous research showing the harmful effects of smoking, there was a master settlement agreement with the tobacco companies.

One of the results of that master settlement agreement was that the Legacy Tobacco Documents Library and other libraries that housed the documents that tobacco companies had, were forced to hand over their research. In this library you can find how R.J. Reynolds, with help from the industry and PR firms, suppressed research about tobacco. They targeted Evangelicals, GOPAC (a Republican state and local political training organization), Citizens Against Government Waste, Manhattan Institute for Public Policy, and the Moonies - the followers of Sung Myung Moon. (Sung Myung Moon, who's now deceased, owned The Washington Times where a lot of global warming denial is still written.)

The Advancement of Sound Science Coalition was set up in 1993 to present a side of the tobacco story in a way that would not be directly connected to the tobacco companies. It established itself as a credible source for reporters when questioning the validity of the scientific health studies. It hid under the façade that it was a grassroots study, people at work, not a study coming from big tobacco corporations.

In order to avoid cynical reporters from major media, these projects were launched in places like Columbus, Ohio and New Mexico. Anything that upheld the position of the tobacco companies was labeled 'sound science', as compared to 'junk science' – which was anything opposed to that position. Its purpose was to defend smoking. These same organizations are involved in disputing global warming. Books and films devoted to telling this story are 'Merchants of Doubt,'(Naomi Oreskes and Eric M. Conway), and 'Dark Money', (Jane Mayer).

> The Orwellian strategy of dismissing research conducted by the scientific community as junk science, and elevating science conducted by product defense specialists to sound science, creates confusion about the very nature of scientific inquiry and undermines the public's confidence in science's ability to address public health and environmental concerns.

In another book, 'Fighting Chance', Arthur Robinson, a Christian Reconstructionist, used the same tactics this time for nuclear war. He wrote about the need to build bomb shelters funded by the federal government. This was the 1980s, and America was still in the Cold War. There was an escalation of weaponry and of building nuclear weapons as a counter to the Soviet Union. In the book, the author insists that nuclear war is not as dangerous as everybody thinks and reinforces the idea that with bunkers for everyone in the United States, we could escalate the war and live through it. He also pushes that certain myths about nuclear war were fraudulent, such as legions of mutant children from increased radiation, tens of millions of cancer victims from increased radiation, nuclear winter, the 'On The Beach' syndrome, or the very idea that the earth would become uninhabitable.

The author wishes you to believe that you have been sold a tale about how dangerous radiation really is. This is an example of the mentality behind a lot of global warming denial and as Robinson just proved, it isn't specific to any one product. Arthur Robinson is still around - he ran for office in 2010 and he is now head of the GOP in Oregon. He continues to spark media attention, this time on global warming. He sells the idea that taking diluted radioactive waste and sprinkle it in oceans, over our heads, or in our water will help inoculate us from disease and counter climate change.

Since the 1980s Greenpeace has tracked where funding is being spent and found it has largely gone to building a permanent infrastructure of what is called free market think tanks, like American Enterprise Institute and The Heritage Foundation (the foundation providing funding for Republican Presidential Candidate Donald Trump), Cato Institute, Atlas Economic Research Foundation, Competitive Enterprise Institute and Heartland Institute. All of these are now quite involved in the global warming denial publicity and many have relied on agencies like Donors Trust, a nonprofit donor-advised fund based in Virginia, to provide their funding. This network has formed something called The Center For Media and Democracy. This group is also the nonprofit that runs ALEC Exposed (American Legislative Exchange Council) that is the part of that network that writes the congressional bills.

The thing about Donors Trust is they can move money in what would almost be considered legal money laundering. Nobody knows where the money comes from. Mother Jones had an article where they called it the Dark Money ATM. The money can't be tracked. What we know about Donors Trust, what we can track, is that Charles Koch, one of the Koch Brothers and co-founder of the CATO Institute, one of the think tanks, donated 8 million dollars to Donors Trust between 2006 and 2010. We also know that Steven Hayward, a board member and treasurer of Donors Capital Fund has been a Fellow with American Enterprise Institute, one of the grandfathers of free market think tanks, as well as the Heritage Foundation.

One of these organizations, Heartland Institute, put up a very controversial billboard - a picture of the Unabomber – that essentially said, 'if you believe in global warming, this somehow associates you with the Unabomber'. Rush Limbaugh, who worked directly for Heritage Foundation, said, *"In my humble opinion folks, if you believe in God then intellectually you cannot believe in man-made global warming. You must be either agnostic or atheistic to believe that man controls something he can't create"*

This leads us to the next part of our story, the David and Goliath religion part. You wouldn't think that the Koch Brothers, who tie for the fourth richest people in the country, and Exxon Mobil (the world's largest publicly traded international oil and gas company), along with the Donors Trust Fund, could become poor little David fighting Goliath. What can possibly be so fearsome and so huge that it would make the Koch Brothers and Exxon Mobile the David in this story? Well, that would be the Antichrist!

Remember the John Birch Society? It was started in 1958, and in the 1960's it was working against civil rights. It died out for a while, but today it's making a comeback, along with their conspiracy theories that are merged with that Antichrist narrative. One of the founding members of the John Birch Society was a man named Fred Koch, who happens to be the father of the Koch Brothers. One of the ways that the John Birch Society has come back in force is through the Tea Party movement. The idea of a grassroots Tea Party movement was developed by Citizens for a Sound Economy long before most of us heard about the Tea Party. Citizen for a Sound Economy was founded by the Koch Brothers. It begins to all fit together.

You can see reflections of all this in textbooks for homeschooling and private school use, first published in 1989 and again in 1999, and still in use today. I live in Pennsylvania where we are now providing public funds to pay for textbooks like this to be used in private schools. Here is a quote from the textbook:

"Book of Revelation makes it clear that one man, the Antichrist, will rise up and take over all governments and economic systems of the world. But instead of this world unification ushering in an age of prosperity and peace, as most globalists believe it will, it will be a time of unimaginable human suffering, it's recorded in God's Word. The antichrist will tightly regulate who may buy and sell and those who oppose him will be killed."

What's interesting about this is that throughout this chapter on anti-globalist, the globalists have nothing to do with corporations. The globalists are the United Nations, those who strive for world peace, and environmentalists, who are trying to build a one-world government.

They also don't look kindly on acceptance of global warming; in fact, they mock it ruthlessly. Here is a quote from a science textbook where the publisher is A. Beka.

> *"Roses are red, violets are blue,*
> *They both grow better with more CO^2."*

And look at what they think about environmental concerns:

> *"Environmental concerns and their effects on the economy are best*
> *left in the hands of private enterprise contrary to the propaganda of many*
> *environmentalist groups; free enterprise capitalism is best for the environment."*

Why the resistance to belief in global warming if climate change fits right into this doomsday apocalyptic scenario? Here is the answer for that. For 150 years this dispensational narrative was the dominant narrative among people who had literal end of time beliefs, but leading up to the year 2000, and since then, we've seen a dramatic shift away from the doomsday narrative to a narrative of dominion. Dominion is the belief that true believers will not escape in 'the rapture', but instead, will take over the earth. The person who is credited with laying the foundation for Dominionism is a man named Rousas Rushdoony. The belief is that the United States will be reconstructed, according to biblical law, and that therefore the movement was called Christian Reconstructionist.

Now why haven't you heard of Rushdoony if he's so important, if he is driving this huge change in apocalyptic belief? One of the reasons is that Rushdoony wrote very openly that in a reconstructed America, people would be stoned. When Jerry Falwell criticized Rushdoony for mandating the death penalty for homosexuals and drunkards, Rushdoony quickly responded, "Oh we're not talking about stoning drunkards." Because of this stoning of homosexuals and adulterers that Rushdoony advocates, people who adhere to Christian Reconstructionist theories don't always want to admit to it.

Now let's talk about biblical economics. Rushdoony and Gary North (who co-wrote Arthur Robinson's book) laid the foundations for what's called Christian Libertarianism. This is the idea that says you shouldn't feed the poor, or have a social safety net, or have any government regulation. In Rushdoony and Gary North's world, God and the bible ordain this order. They say the bible tells you there shouldn't be inheritance tax, or property tax, or a minimum wage. They say, "The bible mandates free market capitalism."

Christian Reconstructionist theories still comprise a fairly small group, although their ideas have permeated much of society. We see this being spread throughout the nondenominational world and the charismatic world. You might recognize the term "The Seven Mountains Mandate" which says that dominion is supposed to take over the seven power centers of society: business, government, family, religion, media, education, and entertainment. So this creates the perfect storm: they have tied together business, politics, and religion.

All of this information is a little depressing, but we've had 'perfect' storms in this country before, and after each storm we rebuild, we regroup, we try to make things better than before. I'm hoping that what I've done here sheds light on these issues and helps to empower us against those who, for a variety of reasons, resist the scientific evidence that global warming is taking place, that it's manmade, and that we need to do something about it."

"Part of the problem is that if you acknowledge that it is real and that we can do something about it, then you are obligated to do it."

BLOOD MONEY©

Hakim Bellamy
Poet Laureate
Albuquerque, New Mexico
(2012-2014)

"Water is the main ingredient in substantially all of our products. It is also a limited resource in many parts of the world, facing unprecedented challenges from overexploitation, increasing pollution, poor management, and climate change. As demand for water continues to increase around the world, and as water becomes scarcer and the quality of available water deteriorates, our system may incur increasing production costs or face capacity constraints which could adversely affect our profitability or net operating revenues in the long run."

– *From the SEC Disclosure*
of Coca-Cola

The blood
Coursing through the veins
Of our mother
Is blue

Water is the main ingredient
In her offspring
In us

We will incur
The costs

Our profitability
Will be obsolete

Our net operating revenue
Will flat line
In the long run

Which is not very long
At all

While we gamble
On the next strike

We are promising ourselves
A stroke

Spend money
We do not have
On terror
We cannot see
But won't spend a dime
On the climate
Bending in our bones

The flood
Ravaging our sinuses
The apocalypse
Brewing in our gut

The future we can touch
Right now

There is a famine in our soul
Wailing el reparto
But we cannot hear it

Singing the song of three sisters
Mourning maize, bean and squash

In a rush to armor borders
Instead of vegetation
We leave our food supply unprotected
As if there is anything else
Worth saving

Not even fava beans
Can adapt to a drought
Of common sense

When will we ask
If we can weather this wealth?

When will we start blaming acts of God
On the inaction of men?

When will we look the storm in the eye
And do more than apologize?

When natural catastrophes in the past 30 years
Have more than doubled globally
Yet have quadrupled in North America, alone
Will we cast stones at Haiti
And keep calling our mother out her name
Until we become Sodom and Gomorrah?

Our beliefs are not enough
Reality is heating up
You can pray
Or you can plan
But, I suggest we do both

What are we waiting for?

Albuquerque to become beachfront property?
For snow packs to become life support instead of a weekend getaway?
For Atlantic City to resemble Atlantis
For wildfires to replace bonfires on the Fourth of July?

For Mother Nature to stand her ground
Bitchslap the coal and electric plants with power outages every summer
Tide the fossil highways with all of her broken water
And force us to evolve

How come we are not born again, even after
Super Storm Sandy baptized Wall Street?

What if water was currency?

Would we spill it from our pockets
And not pick it up?

Would we hoard it
Under our beds
And in our bodies?

Would we save it
For our children's future?

Tell me how much a football field
Of New York City taxicabs under water costs?

Tell me cost of pretending God doesn't exist?

The biggest treehugger known to man
Been talking to us through plagues, floods and burnin' bushes since Noah's kids

Hey, Oklahoma Senator Jim "Climate Change Is a Hoax" let me give you a tip
If you really want more people of color in your voting district, this is it
Cause when the tide rises 4 feet by century's end
And the coastal states with all their immigrants migrate in, you'll get your wish

Your habit is expensive
Your denial
The costliest of sins

The government spending you loathe will lessen
As soon as you let climate change legislation begin

Don't worry about upsetting the publicly traded companies you hope fund your
re-election bid...
Because the truth is green
And the SEC's been requiring them to disclose the Climate Change risk to their
Bottom line since 2010

Your profitability
Is obsolete

All net operating revenue
Flatlines
In the long run

Which is not very long
At all

Sadly,
In the end

All of us
Won't share in the profits
But we will all
Reap the costs.

WE ARE LED TO THE
THINGS WE NEED

There was one other special magic that occurred in DC. I call it magic because the Universe is infinite, and yet here it was – something we were given that we didn't yet know we needed…an unexpected invitation to attend Tom Steyers' news conference at the prestigious DC Press Club.

Mr. Tom Steyers, a billionaire businessman from San Francisco, had gathered a diverse group of environmental, faith and minority leaders to demonstrate the dangers of the Keystone XL Pipeline, an oil pipeline that a foreign corporation, TransCanada, was attempting to force across the United State to transport tar sands from Canada to China. As it was pointed out that morning, it wouldn't even be oil for our country. Mr. Steyers' goal was to insure that President Obama rejected the Canadian project.

Mr. Steyers was joined by Gene Karpinski, League of Conservation Voters; Susan Casey-Lefkowitz, NRDC Action Fund; Leslie Fields, Sierra Club; Van Jones,

Rebuild the Dream; Patrick Carolan, Franciscan Action Network; Rev. Ron Stief, United Church of Christ; Rose Berger, Sojourners Magazine/Sojourners Creation Care Campaign; Former Sen. Timothy E. Wirth; Suzan Shown Harjo, Morning Star Institute; Andrew Nazdin, Power Shift; Rev. Lennox Yearwood, Hip Hop Caucus; Ellen McNulty, National Wildlife Federation; Marianne Comfort, Sisters of Mercy of the Americas; and Conn Nugent, of the The H. John Heinz III Center.

I had never seen tar sands before, and this was one clever demonstration. Rev. Stief took a large jar of the junk and threw it, one spoonful at a time, onto a white canvas to show that it wasn't oil at all. The black splats stuck where they landed. They didn't run down the canvas like spilled oil, they stuck like tar. And to turn it into oil, we learned that the tar had to be treated with at least seven very nasty chemicals; benzene, toluene, ethylbenzene, 1,2,4-trimethylbenzene, xylene, chromium and lead in order to flow through the pipeline. This was my first introduction to the Keystone XL tar sands, but it wasn't to be my last.

At this point in the adventure, one thing was becoming abundantly clear – global warming was being fueled by both ignorance and greed. The cost of coal isn't cheap when one factors in human dying. It may be a freak accident to crush a 3 year old in his crib, but what about the babies born with deformations, autism, and other illnesses because of the pollution created by our demand for cheap solutions? How about the adults with heart attacks or two or three different kinds of cancer because of the toxic chemicals that have settled over towns or poisoned water in the mining communities? These things weren't happening to strangers across the globe, these were American men, women and children, who were dying. Where was our humanity in these tragedies? Have humans ever been humane to each other? If so, when did we lose it, when did we stop? What is greed, or what creates greed? What if we didn't fear 'not having enough'? What if we have been confused about what 'enough' is? Perhaps fear is the evil that keeps 'enough' from ever being enough. And finally the biggest question of all, of course, what is the actual cost of destroying the planet we live on? These questions were burning a hole in my heart.

I could also see that tiny shifts can change everything. Tiny shifts were changing me! Perhaps on the other side of this darknes is the magical message that our best days are ahead of us, starting now, if we can actually look and see how confused we have been, and make a moral and spiritual shift in the way we think about these things, the way we live on Earth.

One question, about how I was going to put the film together, was being answered. 'We Know Not What We Do' would be my love letter, to our Planet, to humanity, and to you, as I began to hear the poems, the music, and the gentle voices that belonged in the film.

OUR NEW YORK STATE OF MIND

On we headed to New York to interview Akio Matsumura, a Japanese diplomat who had spent most of his life working with the United Nations. It was time for us to uncover both the conspiracy and the current cover-up around the nuclear meltdown at the Fukushima Daiichi Nuclear Power Plant. On the flight I remembered the first cover-up story, about those power plants at Fukushima, that Peter Illlym had shared with me months before.

"My whistle blowing days started early, " Peter said during the mountain top removal interview in Appalachia. "My father was a ceramic engineer. He worked for Babcock & Wilcox, and he actually invented insulation that was used in the nuclear industry. Babcock & Wilcox was the company that built the Three Mile Island reactor. They were building a reactor outside the university where I was a student at the time. That reactor was considered defective to the point where my dad and the other two engineers who had designed it, protested and quit publicly to become whistleblowers. I had joined with my dad in that small group. We got arrested and charged with trespassing, but eventually we stopped the building of the next nuclear plant. My father lost his job at Babcock & Wilcox, and I got thrown out of school, but that was the last nuclear power plant ever built in the United States. Six weeks after that, Three Mile Island reactor had the meltdown. What's interesting is that GE took that technology to Japan, and those reactors had the meltdowns when the earthquake and tsunami hit in 2011. The earthquake crippled the reactors because they didn't have a failsafe mechanism to stop the cycle of the meltdown. I live on the west coast and we're now suffering the radioactive fallout, and the things that are washing up from the tsunami."

I had to marvel at how it was all fitting together. Akio Matsumura was someone who had also followed this story. Hopefully he would shed more light on what had happened on March 11, 2011, in Japan.

Akio Matsumura,
Former Special Advisor to the United Nations
Founder of the Global Forum
www.worldbusiness.org

"After some thought, I have come to the conclusion that we live under two sets of law," Akio began, "government law and spiritual law."

I immediately thought that was one of the most important statements I had heard in any of the interviews. Akio had said it simply – off handedly – and we barely had rolled in time to capture it. Just as we did, a loud New York City clanging noise began.

When you are an independent film on a limited budget, finding a quiet place to shoot becomes a challenge. We had actually taken advantage of a wonderful friend who lent us her apartment in the Village as a place to interview. What none of us had counted on was interview day being the day that a side street would be torn up, and loud clanging New York construction noise would begin.

I threw fairy dust over my head as I walked out to talk to the road crew and felt both amazed and relieved when the supervisor agreed to my daring request to "stop the expensive road work" so we could shoot the interview. I returned with a grateful heart, no time to waste, but then another problem occurred; Akio didn't really want to talk about the frightening thing that was happening at the Fukushima Daiichi Nuclear Power Plant. He wanted to talk about how ego keeps people of power apart. It was very interesting, and it told me a lot about the challenge we face in getting authorities to both admit the truth about the dangers, and to work together on necessary solutions. However, an interview about the past wouldn't help the film explain that disaster.

"I established the global forum in 1985," he began again.

My mind was going a mile a minute…1985 was 31 years ago… all that work, and it hadn't done a bit of good that I could see!

"International conflict is created by humans," Akio continued. "That is our basic challenge. My challenge was how to bring political and religious leaders to the table in their individual capacity rather than on the behalf of their government or a church…"

As I listened to him talk about his past, I began to understand how easy it is for us to get caught up in our own stories. Who wouldn't love Akio's story! It was exciting, full of invitations to the Vatican, and even included visits with the Dali Lama, and Mikhail Gorbachev. Akio had indeed put a great meeting together with important leaders and Nobel Laureates. And while I was sure that accomplishment changed Akio forever, other than that it changed nothing. That was twenty years ago...and we now still face the possibility of human extinction from the disaster in Japan.

Finally the conversation turned toward Fukushima...

"On March 11th, 2011, the Fukushima earthquake and tsunami occurred. " Akio began, "At that moment I didn't know anything about nuclear power plants but my friend, a scientist, contacted me. He spoke with great urgency, "Akio, we have a big problem understanding the challenge of the crisis in Japan," he said, "This Fukushima accident might be almost impossible to restore, scientifically. I can only advise you to tell your Prime Minister that this is an unprecedented calamity; the worst humans have ever faced. We need to bring together the wisdom of experts to seek a solution." So of course, feeling panicked, I called it to the attention of Japanese political leaders immediately.

We had discussed many nuclear issues, however, we never talked about how a human accident would affect our health for tens of thousands of years. I had never thought of this. You have to keep radioactive waste for 100,000 years. One accident can cause thousands of years of damage, and change civilization. We have to now come back to the basic question of whether our human technology can match the power of nature. Scientists are stupid if they think they can compete with the power of nature: a mega earthquake, a tsunami. It's a mistake to think we can control the planet.

We are a tiny, tiny species, and we are the newest among species, so we have to understand we are merely one of the tenants of the planet. We started something the wrong way – a mistake that already cost thousands of years for our descendants. The issue is not whether we need energy. The issue is 'what if' the disaster related to that need contaminates our food or even the Pacific Ocean - neither of which can recover for a thousand years?

I would like to explain little bit more about the Fukushima situation. There were four nuclear power plants that were damaged. Reactor number one, two, and three's damage have not been repaired at all since Fukushima because nobody can approach due to high radiation. Of great concern to scientists now is reactor number four, the crippled reactor, still there, a hundred feet above ground in a spent fuel pool containing 1,535 nuclear assemblies. How much cesium 137 radiation is there in the pool? It shocked me. Reactor number four alone is ten times greater than the Chernobyl accident and fourteen thousand times that of Hiroshima's bomb. Tepco, 'Tokyo Electric Company', admits it will take forty to fifty years to safely decommission the nuclear power plant of its radiation. They say forty to fifty years, but my friend, the scientist, says that's not possible. It is a global catastrophe.

It is 100% sure another earthquake will hit Japan before the reactors are decommissioned. So, what do you do to stop further damage? There are two options, really: either the radiation is too strong so we cannot repair it and leave it as is, or, as we have caused the problem, we have to make an effort and a sacrifice in order to save our descendants, and our world. It is evident you cannot mend this kind of an accident without the sacrifice of human beings. There is no question that all living beings will be affected. If even one of the four reactors collapses, no one can approach there within 50 to 100 miles. Whether you wish to save our planet or not - our descendants, our species, animals - what is our moral obligation for the next generation?

And what do we mean by human sacrifice? At the Chernobyl accident, twenty-seven years ago, the Soviet Union, sent between 600,000 to 800,000 military, engineering, coal and oil workers to the rescue. We do not know exactly how many died in order to contain radiation damage immediately, but they took the risk. What is the difference with Japan? If technology cannot fix the radiation then do we not have to sacrifice? If we rely on the development of new technology, nature may not wait. It is a delicate political issue, but it is counterproductive to accuse rather than reveal the basic problem. The Tokyo Olympics, for instance, may raise such an issue. They may send medical or nuclear experts to examine, as it is a moral issue. They are responsible for the risk of participating athletes being exposed to radiation.

Very few people truly understand what is happening at Fukushima. I am not a nuclear expert; I am speaking from instinct. A disaster has occurred and it may get much worse. Unfortunately, human beings are not clever enough to see the long- term effect. Now, after three years, all evidence leans towards the possibility of a mega earthquake.

What can we tell our children, and our grandchildren? That they had a stupid grandfather or grandmother? Why do opinioned leaders in America not recognize this crisis? You 'didn't know' is one thing; you 'didn't want to know' is another. If you don't know, I can teach you. If you don't want to know there is nothing I can do."
No wonder he didn't want to talk about it. Interviewing Akio affected me so deeply because in the entire two hours I couldn't get him to say one positive thing about

Fukushima, yet I had been able – without being a starlet or having a skirt blowing over my head – to experience the kindness of a crew of hard hats to stop their noise so I could do an interview for a movie they would never see. The irony of all that was too strong not to blow my mind.

KIVALINA

Katharine Hayhoe had been the first interviewee to utter the word Kivalina to me. Her question was simply to ask if I had ever heard of it, or the lawsuit that the tiny town was bringing against big oil companies. My answer was no, I had not heard anything about it, and I faked the fact that I couldn't even pronounce it. Katharine pointed out that it was the first town in America to sue the fossil fuel industry because climate change was causing a rise in the ocean and the tiny town – home to 399 Inuit Eskimos – was being washed away. The warming seas were also melting the ice and changing the way the villagers had lived for thousands of years.

Wow, I thought, we have to go. It wasn't only the idea of going 30 miles above the Arctic Circle that excited the heck out of me; Kivalina seemed crucial to the story. Words are great, but there is nothing like taking the eyes of a camera into a place to show instead of tell. This was a place that most people had never seen. It needed to be in the film.

Going above the Arctic Circle was pretty ambitious for a small indie, and it was made much more difficult by the fact that there seemed to be only one phone and one e-mail to use to even reach them for permission to visit. Meanwhile, who would go? I obviously had to be there, and Aaron wanted to go, and Matt and the camera and a lot of equipment had to be carried in.

By that time I had heard an entertainer named Jerry Fairis perform. Jerry is both a poet and songwriter, and I had fallen immediately in love with his down home funky original sounds, and his craggy voice. My heart dictated that he had a place, like a wandering minstrel, in the film, even though I didn't yet know quite how that would work. And of course people were telling me that I had to have 'a celebrity' to narrate the film, but since I felt that celebrity worship in America was partially responsible for why we had wandered so far off track from our humanity, I wondered if I could get away with doing something wildly different. There was also another heated debate: Did we take Jerry with us to Kivalina, to play his guitar on the actual beach that was being washed away? Or did we green screen him?

I laughed when I found out it was going to take three planes and a full day of travel to get us to Kivalina, Alaska. What I didn't realize was that the planes were going to get smaller and smaller. The first stop on the first flight was Anchorage.

After a few hours wait, we boarded the next, slightly smaller plane, and that is when Fate started to take hold. A pretty young blonde who was hurrying behind me actually stopped for a second to ask me if I needed help with my carry-on bag. I said no thanks to her offer of help, and after getting my gear stowed away, I noticed that her assigned seat turned out to be next to Jerry. Interesting, but I still didn't realize how much Divine Grace was playing a hand.

We reached Kotzebue, the last town before Kivalina at about 10am or so, and were disappointed that we couldn't get on the tiny 11:00am plane to Kivalina. We would have to wait for the afternoon plane. We were 'stuck' in what seemed like a tiny town. We later realized that it was a big town compared to Kivalina itself. Meanwhile, Jerry introduced the blonde woman who had been his seatmate. Sarah was an elementary teacher who worked in the Kivalina School. She had also been too late for that morning flight, so we invited her to lunch and knitted together in the way only strangers can, when suddenly thrown un-expectantly together in a unfamiliar place.

As we ate, Sarah told us stories of her experiences in the tiny school. From the start she made our visit easier. She had guided us to the best lunch café. She also knew a Kotzebue taxi driver, and talked him into showing us around the island after lunch. We got to experience walking on the tundra and weren't too surprised when we were told it was drier than normal because of the changes in the climate. Feeling right at home under Sarah's guidance, we then returned to the airport and awaited the tiny plane that would carry us forward.

I don't remember ever being on a plane that small. Everyone had a good laugh, including me, when the man loading the plane commented that there might not be room on the plane for that small pink bag (the one I had been carrying from plane to plane since Albuquerque). In the moment it took to realize it was a joke, I had let out a scream. That final flight in the low foggy sky was both amazing, and a little scary. Then the real joke was on us, because when we landed on the beach, with no concrete run-way, control tower or lights, we found out that the woman who was supposed to be there to guide us wasn't there, and what's more, no one knew when she would return.

We were shown the only cabin available to accommodate us. It was a small barren wooden square with no electricity and no heat. That is when our amazing grace kicked in: Sarah. What if she hadn't sat next to Jerry on the plane? But she had, and she immediately said, "Come with me to the school. I am sure I can arrange for the guys to pay for a room in the school that will have lights, heat, and showers. Elizabeth can stay with my roommate and I in the trailer next to the school."

We will all be forever thankful to Sarah, and Rachael, and the folks in Kivalina who took us in. The tiny 3 square mile village had no paved roads and no cars. Electric golf carts were the only form of transportation in a town of maybe twenty five to forty small shacks, one community store, the school, 3 churches and some enormous rocks that our government had, in their wisdom, placed on the beach (at a cost of 2 million dollars of tax payer's money) to prevent the erosion of the ocean. A solution, we learned, that wasn't working at all.

Jerry Fairis and his expressive guitar was the key that unlocked the diminutive town. For the next few days he would be giving concerts everywhere: in the church, for every grade in the school, and on our very last night, for the entire village in the gym, inviting other town musician to join him. Jerry was really a good lesson for this director. You can't green screen what Jerry Fairis can do.

But it was the children of Kivalina that floored me. They have never been out of that village. They don't see many foreigners. They exhibit absolutely no fear. They have a vibration that is so strong as to almost make you uncomfortable. They run together, in groups of 3 to 5 at a time, and they stand very close to you, and almost unblinkingly, they ask questions. "You're old but you don't have many wrinkles on your face," one said touching my face. I thought about it and realized that my years of protecting my face with creams and anything else I could think of, from both the sun and the cold, had created a different 'older' face than their Inuit women had. The children had immediately noticed and were asking me why my older face wasn't weather beaten like their mothers and grandmothers. Frankly I was astonished. It was the most honest stare and scrutiny I had ever been under. May I never forget the honesty of the children of Kivalina.

Meanwhile Colleen Swan, a former Tribal Administrator who continues to be deeply involved in the relocation of Kivalina, was willing to be interviewed.

Colleen Swan
Former Tribal Administrator
Leader of the Kivalina Committee on relocation

I had begun the interview by asking a question about the terrible problem of the melting ice. She attempted to answer the question, but then sputtering, she stopped and glared at me.

"I just can't do this," she said, "I mean you people come up here, and just go right into asking me to tell you about the ice. I usually associate that with what happened in 2004, or to whaling, and why the ice is important to Kivalina. I mean you could get your information about the ice from anyone. This is not going to work. It's just not going to work. I can't do this any more."

With those words Ms. Swan got up from the couch, unhooked the mic and headed for the door. The 24 hours of travel, the three planes getting smaller and smaller, and most of all the disappointment in myself for being so linear, so insensitive, hit my chest and brought tears to my eyes. With the opening of my heart the 'she and me' disappeared.

I got out of the way and the voice of the Spirit of our oneness spoke. Our wet eyes met without blinking; we seemed to melt into each other. I began to breathe again when she returned to the couch and clipped the mic back on to her blouse.

"I've done so many of these interviews about climate change and how it affects Kivalina," she began again. "It's been an uphill battle, trying to convince people that this is a serious problem. It is affecting our culture. It is affecting our location, and they go together. We depend a lot on what the ocean provides for us. Ice is very important, not just to us in Kivalina, but also to the animals - the sea mammals.

People are still in denial, but no matter what people choose to believe - if they choose to believe or deny that climate change is happening - it's still real to Kivalina. I wish that people would change their ways, but we can't even get our government to take serious action. They come up with band aide solutions (like those rocks) that won't prevent the worst from coming.

Occupy Wall Street was all about how the 1%, the rich, will always get what they want. The rest of us, the 99%, we have to fight against these people who have the power to make the decisions and are not making the right decisions for the climate. They can fool anyone. It surprises me. It really surprises me that people actually believe what the government says. Scientists are starting to realize that their projections were wrong. We're not going to see the effects of ocean acidification in 50 years. It's going to happen in 10 years. It's happening now. It is a serious problem. So we have this battle between the material world, and the spiritual world.

Our own system, our Inuit spirituality, our people's spirituality before Christianity, was very similar to what the Bible teaches. Everything that's associated with our Inuit spirituality is written in the Bible, but we never knew that. I've been looking at our history and how Christianity first came to be in Kivalina. It goes all the way back to the Doctrine of Discovery. You might wonder why Kivalina has Episcopalians and the French church, and why does Barrow have Presbyterians, and Kotzebue have Catholics, and so many other different denominations? They all came to 'save us'.

That had a devastating impact on the identity of our people that still exists today. The elders still remember the things the priests first told them. It's an old system that everybody uses to conquer others. It doesn't matter if they are industrial developers, the government, or an educational institution; they come into villages with the mentality of 'saving us' and making sure we're able to go to heaven, but in reality, their real message is one of control. They come to control the people and take over the land. Why? Because it is rich. The Doctrine of Discovery was all about conquering land that was rich in minerals, and Alaska is very rich.

Our Episcopal church has apologized to the native people for their role, but an apology is not enough. It's a start. They have repudiated the Doctrine of Discovery, but it shouldn't end there. There has to be restitution. There has to be restorative justice to give back to the people what the churches took, stole. Those people, who knew the Bible, yet left certain things out. They chose to withhold things because they meant to control our people. If they had shared everything that the Bible said, it would have given us power. We are changing that now.

I think knowledge is our strongest weapon against all of these influences, to deal with this changing climate. Not just knowledge about what everyone outside brings, but about us, because our people have forgotten what we really look like. I spent 20 years as Tribal Administrator and my job was to protect our people and to preserve our way of life. That was very hard to do because even I couldn't see who we were anymore. I thought our people were poor. We have no cash economy. All these years I looked even upon my parents as very poor people. I believed that our ways were going to die. That we were going to disappear. That we were changing in a way where we wouldn't be able to survive as Inuit people spiritually.

Then I met a whaler from another community. This guy was ashamed of his situation. He was ashamed that he was poor. He lived in a shack, but when he was talking about whaling, this whole different person emerged, someone very proud - a very successful hunter. He reminded me of my grandfather who was a whaler. My grandfather was not ashamed. He was cashless. He was 'poor.' That was the way he lived. Successful whaling captains are the icons of the community. They set examples. They're leaders. They feed the poor. Their freezers are always full. They are rich in other ways. By our standards, my grandfather was the richest man in the village, the most powerful, and the most influential. That's what opened my eyes.

Looking at our culture through civilized, western eyes is designed to make us feel ashamed, but when we look at ourselves through our own eyes, through our Inuit customs, our traditions, our roots, it's a whole different story. The poorest people in this other society, are the richest and most powerful in ours. That is when I started to realize that our people are still alive. We still follow our traditions. We are not gone; we will never be gone.

It says in the Bible to always respect the elders. Our elders are the most respected in our community. We don't put them in senior centers where other people take care of them because we're too busy. We don't do that. They're the people who make the customary laws that we live by - it's what the Bible teaches. I don't know why people don't follow what the bible teaches; it is the best system in the world.

When a young hunter goes out and catches his first seal, it's celebrated, but he has to give it away. I compare that to what the bible says about giving the first fruits of your labor, or tithing, because you get it back. When a captain gives his food away, it always comes back.

When the missionaries came there was a language barrier. They didn't understand our celebrations. Our people spoke only Inuit; they didn't know English. When these missionaries saw our people dancing, they saw it as voodooist or, shamanism - devil worship. Some people actually still call it devil worship here in the community. That's how influential the missionaries were. That's what nearly destroyed our people.

When I was Tribal Administrator I had to get to know our people, our culture, again. I heard an ordained Episcopal Church deacon say, "God speaks to his people in a way that they will understand." That stuck with me, but it took me 23 years to figure out what that message meant. To know for sure that God had never stopped talking to us in the way we understood. And He used the whale to get His message across. I had to experience it in order to be able to get it. I was married to a whaling captain. Our elders told us, "You have to be a good person, and have a clean heart, because when you're out whaling, the whale gives itself to you. That is a huge responsibility. I never knew how big a deal it was 'til we landed a whale and it was the most humbling experience of my life. The spiritual part of all of this is so private that it's not going to mean anything to anyone. You can't possibly know or understand it. I mean, you hear the words, but it is folklore – it's an old wives' tale."

I was stunned with Colleen's story because it was the same story that the Native Americans had told me about hunting Buffalo. "The Natives do ceremony the night before, and the Buffalo 'gives itself to you' on hunting day."

"The way we live falls under the protection of the Freedom of Religions act and it is being violated because of economic development - because of money. The government, the oil industry, and the mining companies are all violating it. Throughout the 20 years as an administrator fighting for our people, I earned a reputation. It was a dirty fight. I was called so many things. One of the most amusing to me was 'disenfranchised militant.' I was in a fight against environmental pollution because our people rely on the food that the environment provides for us. It's our economic base. It caused me to violate one of the strictest customary laws that my grandfather taught me - to avoid conflict. In my fight I was breaking all the rules, ignoring all of our values, but at the same time I still felt guided by God. I really don't like being angry; I don't like fighting, but God made it impossible for me to stop."

And with that Colleen began to cry, and we all cried with her. She had touched our very core. It was quiet as we waited for her to begin again.

"I always tell myself, "You're not going to do this," she said. "I don't like crying in front of the camera. I don't like getting emotional. I still have that fighter mentality, you know, never show weakness."

"Emotion is not a weakness Colleen," I whispered.

"I know it's not but I still don't like to show it. (Now she dries her tears and chuckles) And I always tell people, when they're trying to be strong and trying to hold back and not to show emotion, "Just get it out. Cry it out." But I still have to maintain this image, because I'm still going to have to fight.

Then she went on…the first years of my fight to protect our people began with the relocation project. Our tribal government was being ignored by the regional organizations, and I had to demand attention. I asked the chairman of that committee, "Why are these people avoiding me? If they don't pay attention, I'll fight to make them pay attention." His answer was: "Because they're afraid of you." I asked him why and he told me that 'they are afraid because you speak the bold truth.'

I was naïve at the time. I believed that everyone who had a responsibility to the people told the truth. I was finding the opposite was true - especially with the climate change issue. Our government is taking non-action to address the climate change issue. They look like they're taking action but they're not. One of the parts of our lawsuit against the oil companies, and the energy companies, was the science that the oil companies themselves had paid for. They commissioned false reports. The other part was the pollution, the carbon emissions that caused our problems. We lost that lawsuit because the people who have the power to make decisions say we have to have a point source. It's almost impossible to go to one single source and say this company caused our problems with their carbon emissions.

I really had hoped that we would win that lawsuit, not for the money, people to wake up and pull their heads out of the sand, you know. This is a deadly problem that affects everyone. Not just because of the changing environment, not just because of the situations like ours, but carbon dioxide is also causing the ocean to become very acidic. That is what people say is climate change's evil twin. And it's an even worse problem, because it affects the food chain…the beginning of the food chain."

She stopped now, and smiled, at peace that she had gotten a deep part of the story out to us. "I can take questions now," she laughed. Everyone was quiet as Colleen and I gave each other a long hug. Everyone had been changed. Colleen for having trusted us enough to finally speak from her heart to strangers. The rest of us, because we had come to Kivalina, thirty miles above the Arctic Circle, and found family.

Enoch Adams

Enoch's house was a mansion compared to most of the shacks in the village. We were greeted warmly into a house full of pictures that reminded me of Tug's place. We had been told that Enoch could tell us about the history of the village. I had already learned from Colleen to just let him tell the story.

Enoch Adams
Inuit citizen of Kivalina
Hunter and Family Man

"A group of people came from Washington DC to scope out Kivalina for developing offshore oil," Enoch began. "We were voicing our concerns about how we use the ocean for our food and one of the officials made an off-hand suggestion about just getting supplementary foods brought up here from the lower 48….until somebody questioned if they knew how much it cost for a gallon of milk. They were amazed that a gallon of milk, bought in to our one island store, costs $10!

We don't just use the resources the way we do because we live here and it's our custom. We use them because it's cheaper. It's cheaper to go after our own food. We may dress like and talk like western civilization - we eat some of their foods - but we're still dominated by our culture. Your cultural domination has not overrun us. We're still pretty strong because of how we access our resources here. The official I was appealing to, about off shore drilling, started changing the subject. So I told him, "You know what you need to do. You know how you can safely do it, but it's going to cost a lot. You aren't going to make a profit the way you need to, the way we know you want to."

I also told him what I had heard on TV, what a native minister was saying about our people needing to repent their dependence on western culture. The white western culture has changed us, and a lot of it has to do with the technological explosion that has happened over the last 50 years. Technology has virtually erupted and it's changed us without anybody really noticing how much they've changed.

I went to Jerry Falwell's University - a school that completely denies climate change. I had several conversations with Jerry personally. We agreed while talking that greed has become the most prominent characteristic of western civilization. Even he, a born-again fundamentalist, has an open mind. "Do as little as possible to change who your people are as a culture," he once said. "Jesus did not come to change cultures. He came to improve them." Jesus tells us not to judge one another. We really don't know one person from the next deep inside their hearts. Even a husband and a wife cannot communicate fully what is going on in their hearts, and it's very easy for people to deceive one another. The only one who really knows what's going on in the heart of each individual is God. The gospel is a deeply individual message and it's so personal, but when you see a person change because of the gospel, everything that he knows and does improves. The bad things - hate, malice - we're commanded to put aside or take out of our lives. We're take out greed. We're to take out adultery. And when we do, we're left with is what's beautiful about us.

When the gospel came to us it was easy for our people to gravitate toward it because a lot of the tenants of the Christian doctrine were already a part of our lives. We were taught to share what we get. We were taught to be generous. The family unit is so important to our culture and we've become tight knit in our community. Our parents and our grandparents taught us to take care of our families and make sure there's food on every table. These are cultural values that we practice. We need these things in order not just to survive up here but also to thrive. We've been called the Nuevo Savages; we're used to that title. In reality, we have practiced democracy in our communities. That is how we have governed ourselves. Our leaders are selected by acclamation. That's what our parents taught us, and it has been part of our culture for thousands of years. We live by the same rules that govern most communities.

This environment is very dangerous when you don't live by the rules; even the water in the summertime can kill you because it's so cold you can get hypothermia. We live in a dangerous, vulnerable environment and we have to take care of it. We don't think of ourselves as owners, you know. We don't own the sea. We don't own the land. Ownership comes with property; our boats, our homes, our tools. Everything else belongs to everybody. We're just here, as part of the environment, trying not to do anything bad to it, so it can keep providing the resources that we use."

At this point I was reminding myself that I am up above the Arctic Circle with a member of an Eskimo tribe who is talking to me about how to live in a truly a democratic culture. I was not only humbled, I was quickly realizing that I didn't know anything at all.

"Our resources are renewable," Enoch continued, - "the caribou, the walrus, the beluga, the whale, the berries, and the greens - if we do not cause harm. If we do not overharvest, we know that next year they're going to be there, just the way they've always been for thousands of years. That's a point that the western culture has forgotten. To them, renewable means you find it, extract it, finish it up and go find some more.

When you have a big family, you need to harvest at least four seals in a year to survive well. Minimum is four for our family. There are some families that need ten or more, but the norm is around four. An adult seal weighs between four and six hundred pounds each, and they have a lot of blubber, so they provide a sustainable food source when frozen. Sadly the season for harvest has gotten quite a bit shorter. Before climate change became a factor in our livelihood, it was normal for the season to last up to a month. Ice would be ten to twelve feet thick. There'd be a lot of it in the ocean, so much, that even as it broke up, it would hang around for a month. Now the ice has become so thin, only around a foot or two. That makes it very unstable in providing a platform for the seals. The last few years, we've been lucky to have the ice stay for a week.

It really started to become an issue for us in 1998. We were able to hunt for only three days. The ice broke up and when we started hunting…not knowing that it was going to leave, we hunted as normal. We went home, rested, and waited for good weather but by that time, the ice was all gone. There was no more ice in the ocean in the spring. What normally stayed for a month was gone in three days. Some of us were lucky enough to get two or three seal - some families didn't get any at all.

Seals follow the ice. When the ice is melted, they follow the ice edge as far north as they can go. The spotted seals mainly eat fish so they hang around inlets, and channels, or around rivers. The others follow the ice. We don't harvest them in the fall because they're way out there, making them much harder to find.

Now that the ice is gone, we have to hunt the entire time, without rest or sleep. We have to go out as far as possible and stay out as long as possible, until we get enough. I know we'll be fine when I get my fourth one. Of course, we have to wait on weather conditions. You can't just go out there when it's bad. It's dangerous. You can die out there if you don't follow what you've been taught. You learn about those danger zones. You avoid them. The way that we're taught is by example. Sometimes there might be loose ice under an iceberg that could come up and destroy your boat, or wind and ocean current could close up and cause you to be stranded.

Another source of food is caribou. One caribou, if you eat it for lunch and dinner, could last about four days for a family of six to ten. Caribou always follow the wind. When the caribou are not there, you know something is wrong ... and they have been missing since the climate changed. In summertime there's mosquitos. So many that they can affect caribou numbers and the way the caribou deal with it is by running towards the wind. The last few years, when climate change or global warming first touched us, one of the first things that we noticed was the prevailing winds had changed direction. Before the prevailing winds were from the north, but now, especially the last ten years, they are from the west. In the summer time, before climate change, when the wind blew from the north, it was always clear; no clouds. The last few years, the north winds have been bringing rain and that is a really weird thing for us. The north wind? Rain? Something's wrong. Something big has changed, and it has been a huge factor on how the caribou migrate. Let's call it what it is: climate change, global warming.

Everybody will have to change; stop burning so much coal, stop consuming so much oil. That's the only thing that'll slow this down. Then maybe we would be able to get back to what used to be. I don't know if we can get back to what was normal, but there's always hope. Please; get your head out of the sand. Our eyes have opened, those of us here who have to deal with climate change. We've experienced what is still just a theory for a lot of people. For us it's a fact; we live with it, and we've had to adjust to it, drastically. We have to monitor the kind of changes that have taken place so we will be able to harvest the next year. We've had to change what we do, to learn and relearn, our harvesting methods. How did it change? What took place? Is it going to be that way the next year? You become acutely aware of your observations over a period of time, not just a season but from year to year. You have to change because what used to be is not there anymore.

That's the kind of thing the rest of the people in the world will have to do. They need to see what is happening up here and make changes now if they don't want to go through what we're going through. If they don't change now, they will be forced to change later. I'm hoping they will become aware of what they need to do so that they won't have to experience what I'm experiencing. Right now they need to begin to wean themselves off of coal, oil and gasoline. Only the rest of the world can determine whether or not this climate change is going to be worldwide. It's coming. It's moving southward toward the equator.

We started experiencing climate change about ten to fifteen years ago. Then we started noticing the changes had moved to about 100 miles south of us. Then it was about another 150 miles further south. I'm sure that it's doing the same thing reversely, from the South Pole; it's moving northward towards the equator. Climate change is moving and when it reaches the equator, it will become worldwide. People are going to look back and say, "Man, maybe we should have listened."

When something out of the ordinary or unusual occurs in a season, you treat it as an anomaly because next year you notice it doesn't happen again. But when it happens again, year after year, you have to change. Kivalina may be under water within the next ten years - I hope not, but the science seems to point toward that. The best thing we can do is try and make people aware so they can change before the same thing happens to them. If Kivalina goes under water, a lot of other cities and communities along coasts around the world will be under water too. New York will be under water, so will Los Angeles, San Francisco…Tokyo…even London. The severity and vulnerability of living here makes it the obvious early warning.

We're the canary in the coalmine. Really. I mean, when you really think about it, that's what we are. If you don't do anything now, the same things are going to happen to you. I mean that's the whole purpose of the canary in the coalmine. You know? To make changes. Wait. Stop what you're doing. And then, hopefully, everything will be okay. That's the purpose of early warnings. But to listen is the difficult choice that the rest of the world has to make.

There is enough energy in this world that we don't have to depend on coal and oil. People can get off their chairs and get to work and still be comfortable. In order to do that though, there are some things we have to stop - not everything, just some things. The goal is to leave as small a footprint as we can so the next generation can enjoy the same things we enjoyed. We don't need a large footprint. There is no need for us to be greedy. Technological advances have changed everybody. Even the western culture has changed quite a bit and a lot of them don't know it because technology is the mainstay of their culture. The last fifty years has done a number on everybody. Everybody. Not just western culture. At the discrimination conference I asked what the most consistent thing about man was. The answer was change. Then I asked them what the one thing people resist the most was. Same answer. Everybody has the capacity to change but they don't want to.

Have you ever looked at the true definition of the word 'repent'? It doesn't mean you aren't going to sin anymore. Repentance simply means a changing of the mind about oneself and what God has told one to do."

After Enoch's interview, he had his mom show us how she cleaned the seal's skin and made shoes for the family or to sell to others. She spoke of many of the seals being sick, now, and bleeding from their eyes. It was a sad commentary on their changing world. But there were moments of laughter. We will remember forever her telling us, in all seriousness, that the perfect temperature for the outdoor work was 40 degrees below. Then she smiled and added, "but I don't go out to work when it reaches 60 below, that's too cold."

On our last night, Jerry did a concert for the town and Enoch came, guitar in hand, to join him. When it was time to say a final goodbye, I was so touched that I cried. "I will never forget you – I love you," were the words I blurted out. He smiled. He knew I was talking to everyone in Kivalina.

Kivalina Postscript

During our very last day, there was one scene that was unfortunately burned into my brain. After hearing Colleen's comment, that her "people had lost the ability to see themselves through their own eyes", I found it a little disconcerting that there was a large flat screen TV in every small house I entered.

However, nothing quite chilled me like the moment a woman took us into her house to taste seal oil. While Matt was busy dipping the frozen fish into the oil, I happened to glance into the one other room – a small bedroom – and saw a young woman, perhaps in her twenties, sitting on the bed with her baby. The baby, obviously under one year, was lying on the bed holding his own bottle as he nursed. The young woman was sitting a couple of feet away, holding her iPhone and texting. Enoch's words roared through my head. "Technology has changed us without anybody really noticing how much we've all changed."

Sadly, that is true of all of us, Enoch…

Hop On The Wrong Plane To Make The Right Connection

By this point, my perceptions about climate change, about the mountains of Appalachia, scientists, evangelicals, soldiers and Eskimos had all been smashed. I had mostly been wrong about every stereotype I had about people and my planet.

I wrote a book in 2013 entitled 'Gifts of Gratitude', published by Burman Books. It was a collage of stories about things that happened to me when I carried gratitude in my heart and didn't get bummed when life took sharp and unexpected turns. I was particularly grateful for that mental training/meditation when my producers, Aaron Daniel Taylor and his wife Rhiannon, and I took a business trip to LA.

We were headed back to Santa Fe when this new adventure began. We had taken an early shuttle to LAX. It was a trip that I could make with eyes closed at this point: get off the shuttle, get through security, wait a bit, board a plane, land in Albuquerque and drive to Santa Fe – all in three hours tops. But this morning, Divine Grace had another plan.

It was not until I unzipped my black computer bag at security and saw the men's shoes where my computer normally rested that I realized, first in confusion, then in certainty, that I had managed to take the wrong bag off the shuttle. I was shocked having never done that before.

My producers asked if I minded if they went on home; they had a baby sitter and two children waiting. So there I was, with a stranger's bag in the airport – mostly frantic about all the notes on the film that were stored in my missing computer. I found myself in a situation that could have left me in a pouty mood, and instead, I was curious about what was really unfolding.

First it took several hours to track down my bag and get it back from the shuttle. The biggest surprise was that the shuttle driver wouldn't then take the wrong bag back. The driver explained that no one had called about it, so they didn't care about it. That seemed odd, but there it was. I had checked the bag for information about the owner, but there wasn't a name or a phone number to call. Here is the funny part: You know all those times you have been in an airport and heard the announcement that if you leave your bag unattended the airport security police will take it? Well, don't believe it! Because the next few hours were spent trying to give the bag to any official that would take it. To my amazement, no one wanted it; not the security guards, not the airline, not lost and found baggage claim. I couldn't try to secure a ticket home until I got rid of the stranger's bag, and I was doing my very best to not succumb to sheer frustration. Finally, surrounded by guards, I got rid of it – without being taken off to jail – then headed back to the airline to get a ticket to New Mexico.

We had headed to the airport at five or six in the morning; it was now nearing noon. There were no more available non-stop flights to Albuquerque. All they could offer me was a flight to Phoenix, Arizona with a two-hour wait for a plane that would take me home. They could have offered me a hot air balloon at this point and I would have taken it…

I landed in Phoenix, had some lunch, and headed to the gate to wait for the plane to New Mexico. I sure couldn't miss that plane; I had gotten the last seat available. It was then, and only then that I finally pulled the book that I had been reading, Eaarth by Bill McKibben, out of my now treasured computer bag, and settled down for the last boarding group to be called. Almost immediately a man walked up to me and asked me what I thought of the book.

It was now two in the afternoon. Look at all the mysterious twists and turns of the day that had to take place to get me to that exact spot at that exact moment. Even then I could have been sulky, and missed the connection, but instead I responded with laughter, "Well," I relied, "it's scary as hell, but it's required reading because I am making a film on climate change."

"Oh," John Atkinson retorted as he handed me his card, "Well, if you are making a movie about climate change it needs to include our fight against the Keystone XL Pipeline in Nebraska."

There it was, the gift of Grace.

The dye was cast. The XL Keystone would be next. John was amazing in both inviting me to Nebraska, and making sure my trip was a complete success. He guided me to interview the best people. He lined up a car and driver, a young activist named Chelsea Johnson, who did much more than drive us from point A to B. Chelsea worked with

John to give me the background needed for each of the interviews. Everyone loved Chelsea, and really, it was she who it possible for 3 camera carrying strangers to walk into private homes and offices. Without Chelsea it would have been too outrageous for staunch Nebraskan's to talk to us at all. The interviews were everything we could have hoped for, with one of the most moving interviews being with John himself.

By the time the trip to Nebraska actually rolled around it was January and the temperature was 26 degrees below zero. It occurred to me to chicken out. Thank goodness I didn't.

John Atkinson
Chief Organizer for EnergyLinc

"The core of what I do is research solutions to the energy crisis, the global warming crisis, and to try to gather people together to convince elected officials, old and new, to do something about the crisis," he began. "That's the center of the bulls eye. We work with a lot of other organizations here in NE, and I don't think there's a secret ingredient about it; working together comes pretty naturally to the people of Nebraska. We're the only 100% public owned power state. There's a lot in the culture that goes towards working together - it's the Nebraska Way - being nice and polite but also working together for the common good, not that we always agree on what the common good means.

Global warming is causing climate change. This really became very evident in 2012 here in NE just like much of the world. We've had wildfires. We didn't get the media coverage because we didn't burn down a lot of 'multi-million dollar homes', but we actually had more acres burn than what burned in Colorado. We've had Aspens that were literally thousands of years old that were killed. The ground was burned so deeply that they will probably never come back. We have had a tremendous toll taken on agriculture with both the cattle and the corn, and that of course is a huge part of the economy of our state. In fact, we would be in a real disaster situation, if it were not for the Ogallala Aquifer, which allows much of our state to farm using irrigated methods.

So we feel like the Keystone Pipeline is a crisis because it threatens the very core of the way of life here, which is based on both water and agriculture. Even though most people are not working on farms, many people grew up on a farm. Many people have grandparents and aunts and uncles on farms. So it's very much woven into the culture here, and farms are very much under threat from all the 'tweaks' that global warming has made to our climate.

What global warming has done is taken the climate that has been beneficial to us and steadily raised it until it has become something new and different and outside the realm of our experience in agriculture and every other way. It is truly a crisis, and it is here.

At the University of Nebraska, Lincoln, they do a poll every year about the point of view of rural Nebraskan's. And in that area, in this so-called red state, 59% of the population agrees that we need to curb greenhouse gas pollution at power plants. Now that is radically different from the point of view our elected officials have delivered. So that's one of the huge gaps we find, and we find it all over the country, that there is a huge gap between what people know, what they feel, what they want and what their political system is able to deliver.

The other day we were doing a cleanup of the office and we unearthed a 2005 issue of National Geographic magazine; it was all about global warming. I've had other people talk to me about TV shows they saw on some cable channel in that same time frame. People get it. They have a basic understanding that global warming is happening. They know for the most part that people are causing it and they want to do something about it.

It is really only the volume of the deniers, the so-called skeptics, and the platform given to the deniers by so much of our media, that has made it seem like there's a debate in the scientific community. It's not a dispute among the people who know what they're talking about. And so people, especially with the Internet now, are able to go out and do their own research and sort through all the garbage. When that switch is flipped, their heart is turned on. They say, "I'm not sure what I'm going to do, but I have to do something."

One of the key missing things is people not knowing precisely what to do. It's easy to say, "Well, cut greenhouse gases by 80% by 2035 and to zero before the middle of the century," but how do you get your local utility to stop making electricity by burning coal? How do you stop them from doing the 'business as usual' step of replacing coal with natural gas? The main component of natural gas is methane, CH_4. It's a much more powerful gas than CO_2, carbon dioxide. Going to natural gas in a system where there's a lot of leakage of methane is extremely dangerous. That's the next crisis we have to avert within the realm of the electricity generation. Luckily, here in Nebraska, we are the third most potentially powerful producer of wind energy. We have tremendous wind resources. If we were to develop that and share that energy, we could make a big difference.

We must make a 180-degree turn. We must stop producing greenhouse gases. It's like the old rule of holes. If you're in a hole, stop digging. We're in a deep hole and to stop that digging – into a deeper grave to be blunt - that's going to take things in the realm of politics and political economy. So the challenge is not only a challenge in technology; we have every bit of technology to solve this problem. The obstacle is the lack of political ability and economic organization to make the change.

The lead-time we have blown over the last 30 years that we should have been in rapid action is something we can't recover; so we have to make up for it. I say 30 years for a reason. In June of 1988, there was a young climate scientist by the name of James E. Hansen who went to Congress and declared global warming a serious problem based on his scientific work that had been published in the 1970's. As a theoretical idea, there was concern about it from the 1930's on. In fact, it was discussed in the Lyndon Johnson White House in the 1960's. The basic science goes back to the middle of the 1800's, that carbon dioxide is not all together transparent to infrared heat radiation. So, by the time you got to 1988, and you had Jim Hansen sitting in Congress saying, "We better do something about it", you immediately had this immense roar of objection funded by the fossil fuel companies, ExxonMobil and other oil companies, and the coal companies.

Now we know that the industrial barons of the industries, the Koch brothers, are behind a lot of the funding; billions and billions of dollars they pay to create a cloud of dis-information. They are, to use the title of that book, Merchants of Doubt. They literally have used some of the same PR firms that big tobacco companies used to convince us that we didn't really need to regulate cigarette smoking, 'because you can't really prove cancer kills anybody'. Well, baloney.

It killed hundreds of thousands of Americans. But we were slow dealing with tobacco because of their immense campaign of disinformation.

Those same people are doing it for big oil and coal and natural gas producers today. And they've created a lot of doubt. That's all they have to do. Taking 1988's testimony as a point of reference, we have been deflected from that time on. Action has been deferred. What we've needed to do has been delayed. Now we're finding ourselves right on the edge where we know that, by the end of this decade, the world's climate will start to tip into territory where it has never been before.

One thing that is happening, by producing a lot of carbon dioxide and putting it into the atmosphere as our waste dump, is that the oceans are also absorbing it. We live on land so we don't usually remember this, but most of the planet is water. CO_2 changes the alkalinity, the pH. That pH value has been pretty stable and that's important. It's important for corals. It's important for all the little critters that make tiny little calcium shells - not just muscles and oysters and clams but the tiny microscopic critters. They are the very basis of the food chain and oxygen production for the planet. What has become apparent is that we are just entering the point at which we've changed the chemistry of the sea surface around the world - just enough to where some of these critters are finding it difficult to make their shells. The pH values are now continuously outside the bounds of the measures that we have had for over a hundred year time period.

If we act rapidly we can delay destruction by about 20 years and give ourselves more time to prepare for the different climate. If we are able to shut off this greenhouse pollution rapidly and soon, then we may buy more time. Perhaps we will not be stuck in that new and frightening world forever. That is a big challenge.

It's why I continue to insist that we cannot give up; it's not too late. But if we wait much longer, it may be too late.

Water is, of course, incredibly important. It doesn't just come out of a tap; it comes out of rivers, it comes out of springs, and it comes out of aquifers. Aquifers are big underground lakes that are not just water, but rocks and soil and clay and limestone. There's a huge pool of it that stretches from here in Nebraska down into Texas. It's not all one continuous body of water, but it is, as a geological formation, essential, and defines much of the character of the Great Plains. It is what makes it possible for farmers in years like 2012 when there's an intense drought, to keep farming.

So here comes this multi-national corporation, TransCanada, and it wants to carve a scar across our state directly through the most important part of that aquifer. In the couple of years they've been pushing for this Keystone XL pipeline, you've seen a number of accidents. Pipelines are not perfect. Just the recent rail accidents show the potential for destruction. In Kalamazoo River in Michigan, there was an Enbridge company oil pipeline rupture that was two and half, three years ago now. They still are not able to clean that up.

We learned some interesting things from that, and also from the rupture in Mayflower, Arkansas, of another Exxon tar sand pipeline. We've learned how to answer some of the quips from the supporters of the pipeline. There was a geologist here in Lincoln, Nebraska, who said, "Well, oil floats. It's not going to go down into the aquifer and damage anything permanently even if there is a spill." In reality, you know that is not true because they're still getting balls of diluted bitumen from the tar sand spill they are pumping out of the Kalamazoo River.

I think many Americans are familiar with the fact that during the BP blowout in the Gulf of Mexico in 2010, you saw oil on the surface of the water that the oil company tried to hide by spraying poisonous chemical dispersants on top of it to make it go away. There are still occasional tar balls washing up from the bottom of the Gulf of Mexico.

In the mountains where I was raised we used to get excited over an artesian spring or well just popping out of the rock, on the mountainside, because the contours of the underlying rock were exposed and you could see the water. Likewise with this aquifer on the Great Plains, there are places you can take a length of pipe about the size of my arm, pound it into the ground and you'll get water out of it. That's one of the things that makes this land so valuable.

If you fly over our crops in a plane you will see circles. That center pivot that makes the circle with its long arms is for irrigation, to spread the water onto the crops. You can't do that if there's not water directly underneath that well to draw from. So when TransCanada came and wanted to dig a 4 foot trench across the state, through the sand hills and put that incredibly poisonous crap from the tar sands through that pipeline, people were immediately alarmed and enraged because it's an insult of tremendous degree.

What would actually be in that pipeline is not just the tar sands sludge, but also bitumen, which is a well-known human carcinogen. The very thought of that poisonous brew spilling into one of the most important aquifers in the world is unsettling and, of course, the company acted in an arrogant, dismissive, and contemptuous way towards the people who were farming that land. They came in and said, "Sign these papers. We'll give you this amount of money. And, you know what? We're going to take it anyway because we can get eminent domain from the court and we're a big corporation with lots of money and a thousand more lawyers then you could ever dream of!"

The result was the beginning of people acting together because, not only was their livelihood threatened, and their homes, but they had just been told by a multi-national foreign corporation that they were going to get steamrollered and they didn't have any choice because that big corporation had all the money and all the lawyers. That did not sit well with the independent farmers and ranchers of Nebraska.

Now you see very conservative Republican landowners standing firm and together, with people of a wide variety of different political beliefs, religions, and philosophical outlooks. They are saying, "No, we will not buckle under; we will not touch our knee to the ground for your economic royalty. This is America. Our ancestors are the people who stood up in the beginning to the impossible task of taking on the world's largest, most well armed empire to wrestle our freedom. We have corrected our own course, abolishing slavery, and expanding our democracy by giving the vote to women. So with that kind of determination, over and over again, we have done the impossible." When people started getting together to oppose this Keystone XL pipeline, the common wisdom was, "Well we may slow 'em down a little, but there's no way we're going to win." But we've been winning for five years now…we have done the impossible.

Up to this point we are undefeated. We have stopped TransCanada. We have stopped that pipeline. We have stopped not only the multi-national corporation, but we have prevented the political people in our state who have betrayed us, from finishing that job. And that gives me hope, to be able to say that. When the situation is dire and urgent, people can come together and do the impossible. It's not only a history lesson. It's not just the American Revolution and the Civil War and the union movement and all these things that are part of our history. This is today. And if we believe we can do it, if we have the gumption and shared backbone and lock arms, we can do it because that's in our nature, which we prove over and over again whenever there is a crisis.

My sense is that there's a large shifting of attitudes that does not see a wall between human beings and rest of the universe. That idea, that we are totally different and unique, totally separate from the rest of the natural world, no longer stands. We do not have dominion. The interdependent web of life is something that we all rely on.

That is the point of view that helps us build determination that we can preserve this world, because it means we will preserve ourselves. To preserve ourselves, we cannot destroy the world. It's pretty much that simple and I sense that many people are coming to that understanding from a lot of different directions. There are Christians who call it creation care. Other philosophies have a different term for it. But I think it all leads to the same place.

Make no mistake, however, we have not yet changed course. We are still on the path of business as usual. There's a little change here and there in how much greenhouse gas pollution we put out, but for the most part, we're very much on that interstate highway at high speed, going exactly in the wrong direction at precisely the time we need to be throwing on the brakes.

I have four grandkids. And I can picture those grandkids... and at some point... they should be saying to our generation, "How did you let this happen? Why can we only read about... the lushness and the beauty and the diversity and the coolness and the stability and, knowing what we can expect.... why is that gone? Why didn't you stop this? What did you do?"

James E. Hanson
Ted Talks 2012
Director of Climate Science, Columbia University
Former Director NASA Space Studies

"Imagine, if you will, a giant asteroid on a direct collision course with the Earth.
This is the equivalent of what we face now.

- Heat trapping gases that will disintegrate the Artic ice sheets.
- Sea levels will rise and destroy coastal cities.
- Temperatures will become intolerable.
- 20 to 50 percent of the planet's species will be driven into extinction.
- Civilization will be at risk.

Yet we dither, taking no action to avert the asteroid.
Even though the longer we wait, the more difficult it will become.

If we had started in 2005 it would have required an omission
reduction of 3% per year to restore planetary energy
and balance the climate in this century.

If we start next year it is 6% per year.

A decade from now it will be 15%, extremely
expensive, perhaps even impossible.

But we are not even starting."

THE KNOCK ON THE DOOR

John made many great interviews possible for us over the next four days. Here is the second one, an extraordinary local farmer whose entire farm was being threatened by the Keystone.

Art Tanderup.
Farmer
Retired teacher
Protector of the land.

"I'm a retired schoolteacher who has moved back to a small farm that my wife's grandfather, started back in the early 1900's," Art began. "We grow corn, soybeans and rye. I do all no-till farming. We try and conserve as much as we can as far as the soil is concerned and help the environment that way.

No-till farming is essentially not using tillage implements, like a plow or a disk, to cultivate the soil. You allow the soil to take in the nutrients that are on top. We plant a cover crop of rye every year over our soybean stubble, which allows the roots to go down into the soil to actually help with the tillage and, also to retain nutrients that are in the soil to protect them. It's one of those things where you allow the microorganisms and the worms and the things in the soil to actually do the cultivation for you. It's much friendlier to the earth because the soil doesn't blow, or erode if you have heavy rains.

Both my wife and I grew up in these areas so we know how fragile the soils are here. For years about the only practice that people did was to plant trees to stop the wind from blowing away the sand. As a former educator, I did a lot of research and came across the no-till farming concept. The Sand Hills are basically an area of Nebraska that has fine, sandy soil. It's very difficult to establish vegetation here. A large portion of the Sand Hills is still native prairie. And the ranchers have to be very careful that something isn't dug up because reestablishing the natural environment, the grasses and so forth over that area, is almost impossible to do.

We got started in this Keystone XL fight in 2012 when we got a knock on the door from a TransCanada rep. That's when we started learning the facts about what was really happening here. We had been busy and had not paid much attention to what had been going on with the Keystone. It had first been planned to be east of us about 50 miles. It was oil, and we just accepted it as most people did. And then the issue came up where the route was crossing the Sand Hills. As a child I'd spent some time up in that area, so I knew it fairly well, so even while the bystander was still being told about all 'the wonderful things' that the oil line was going to do - the jobs, our oil independence. Wonderful things. We said, "Well, yeah, but this is not a good place to build it." And then, Nebraskan Governor Heineman actually agreed and said, "We will not allow that oil line to be built in our Sand Hills or over our Ogallala aquifer. Pick another route."

So TransCanada decided that they would change the route and most people honestly thought they would run that route about 50 miles east of us. Instead they chose to still go through the Sand Hills. They used a map that basically said it wasn't the Sand Hills anymore, but every other map said it was. Even after they came out with this new map, they were still over the aquifer.

Before TransCanada approached us, we really didn't have an opinion on this whole thing. We were still under the impression it was oil and all these good things were going to happen. We had no idea how they mined the stuff. We had no idea what was in it. We had no idea how devastating it was or what it has already done to people; the affects it was having on the climate. We had no idea they were going to run it right through our farm. Our house sits right over where they would put that pipeline through. Consequently, they were nice enough to say, "No, you don't have to give up your house. We'll bend the pipe a little bit and go around your house." They offered to make four bends in the pipeline in a mile and a half. Well anybody that knows even a little bit about Science knows that when you take something and stretch it out, then kink it up a little bit, you're going to make that wall thinner. Any place there's a kink, there's potential for a leak.

Our house water well is about 70 foot deep. My irrigation well is 120 foot deep and it produces a lot of water. It's extremely precious water, quite probably the best water in the United States. In addition, we have very fine sand that turns into gravel, as you get down to that level. The well drillers ran into the shale layers.

Consequently, if that oil line were to leak - and it would - the chemicals would leak down into the aquifer and the contamination could never be cleaned up. The new location of this would be about 600 feet to the west of where we are right now. The chemicals would be pulled into our drinking water and also into our irrigation water. They would be damaging to the crops as well as to any livestock or humans that consumed it. From what we understand, some of the chemicals might be tasteless and you might be drinking this stuff for some time before you realized it was in your water. And with the pipes underground, it could be leaking a long time before anyone would even be aware of it. TransCanada has admitted that they can lose 2% of the volume in the line before it sets a red flag on their equipment that tells them there's oil disappearing. That is a lot of chemicals that would be getting out into the soil. Then the cancers, and other diseases would follow.

In addition to all of this, the Keystone XL is a 36-inch pipe that they are planning on burying approximately four foot underground. It's about 1,500 lbs. per square inch of pressure that creates heat of up to 150 – 160 degrees and would obviously heat the ground above it. It's questionable whether you would even be able to raise a crop over the top of it because of the heat. One of the issues that have come up is the increase in insects that are over-wintering in that warm area. Rather than going through their natural cycles, they are multiplying and farmers are discovering they have all kinds of insect problems that they never used to have.

In Nebraska, the eminent domain issue is a big issue. If the Dept. of Roads said, "We need to widen this highway a little bit, you know, it's not safe. The ditches aren't as big as we need —we need to add another six foot to the width of the highway," you could understand that. You might not like it, but you would say, "Yes, we need to do this." Or if a power company needed to put a power line across your property, you could understand that need. But, can a foreign corporation come in and take a private American citizen's land away? To me that is totally un-American. That should not be happening. TransCanada came in and finagled our state legislature to pass a law allowing them to have eminent domain, so now we have a lawsuit.

There is a theory that the reason they want to come through this area with the Keystone XL is that when the tar sands dry up, and they will eventually if they keep digging them at the rate they are, they will still have easements on all this land over the Ogallala aquifer. If they keep polluting the planet, clean water is going to be worth more than any gallon of oil was ever worth. If that pipeline were across the aquifer, it would be very easy to tap into that supply and use it for corporate greed. They will make their money off the tar sands; and then they will make their money off our water.

Our prayers every day are that the Keystone won't be built. We know there is so much corporate greed and so much lobbying and corporate manipulation and so forth. We've seen a lot of that already and it continues. We feel like we're David battling Goliath, but we know that David was able to win that battle and we feel we can somehow win this battle too. We want to protect this land. We want to protect our water. We want to keep this land as God gave it to us and pass it on to future generations. It's something that is so powerful. If we can put a stop to this, these large corporations are going get the message that you can't mess with the people. God gave us this earth and we're here to take care of it.

This pipeline thing is a disaster but we're so thankful everyday for the community of people and groups and organizations that have come together. It doesn't matter whether you're a Democrat, or a Republican, an ultra-conservative or a liberal. It doesn't matter what your political background is. It doesn't matter what your religion is. It doesn't matter if you're Native American; it doesn't matter. It has brought people together. We come together, and sit down, and talk about an issue that is common to every one of us. And it's become, "We are doing this. We have worked together to build a coalition of people, groups, and organizations that simply says this is not good for mankind. It's not good for our country. It's not good for anyone." That is the most remarkable thing out of this. I've talked to some people and they've even said things like, "You know, we found out so much about our neighbors. We found out who the greedy people are and who the real people are. We found out who really believes in being good stewards of our planet… And we found out those that are ready to sell off for the mighty dollar."

We traveled to the Climate Change Rally in DC last year. We testified at the State Department hearing at Grand Island. I've been working at the county level, trying to get some zoning regulations in. We've been doing a lot of different things. The people, groups and the organizations in Nebraska have been fantastic in stopping the Keystone XL Pipeline. There is no other state that has been able to do what we have been able to do here.

In November we hosted the Ponca Trail of Tears Spiritual Camp here at this farm. And it was so exciting to see members of eight to ten Native American tribes come together and work together on a common goal. For many of these, it was maybe a first interaction with another tribe. For the Ponca, corn is more sacred than tobacco, so when someone asked if we could plant sacred corn with seeds that have been preserved for a hundred or so years, right in the path of the Keystone, I said, "of course we can." So we have hand planted four acres of sacred corn in the Keystone's path, and we have registered these four acres and sent the paper work all the way to Washington D.C. By planting it here the ground becomes sacred ground for a sacred crop, and consequently the freedom of religion that is given to all Americans applies here, and no pipeline can come through here now because of this sacred corn."

I learned a lot the morning we met Art Tanderup and his wife Helen. I had no idea at the time how much I would come to love and respect the two of them. How grateful I would be to sit at their table and share a fresh cooked farm meal. I had no idea at the time that traveling around Nebraska with Art would result in him teaching me how to listen to the voice of the land or that I would learn to see the landscape in a totally different way. It wasn't just the lesson of the makeup of the soil; I was learning to hear the beating heart of the Heartland.

MARY PIPER

It was snowing heavily on the early morning when we went to author Mary Piper's place. I knew she was a 'New York bestseller' and would be an important voice for the film, so I pressed John to get her to agree to an interview. I was in sweats, Uggs, with no make-up when I arrived. She took one look at me and said, "Oh thank goodness, you are one of us." I was thrilled.

I had read her book, 'The Green Boat', and realized how articulate she was in sharing the story of the citizens of Nebraska who bravely took on a fight they were quite sure they couldn't win. I loved her forward of the book that stated she had written as fast as she could when she realized the time line of what needed to be known about the problem. I was experiencing that pressure myself. Mary Piper is a woman I will always admire and for whom I am forever indebted as an important spokeswoman for all that is holy about our Earth.

Mary Piper
Author of The Green Boat: Reviving Ourselves in our Capsized Culture
The New York Times Best Seller list
Lincoln, Nebraska

"I'm a writer, and a community organizer," Mary began, "We started organizing the fight against Keystone XL Pipeline about three and a half years ago. It started when a wonderful friend, Brad, an organic farmer who was good at pruning trees, came over to help me with some yard work. We spent a lot of time sitting outside weeding and talking about the state of the world. It was Brad who said, "You know, we have to do more about all this. We care about it so much. Let's get some people together and talk about what we can do."

So we planned a little dinner party here at my house. The first time, I forget how many people came, three or four maybe. I had already decided that I hate meetings. I only want to work with people I really enjoy and feel are positive and fun to work with; those who get on well with other people. So we had a really nice dinner and some French wine and we talked for several hours. The thing that I noticed about this meeting was that afterwards, for the first time, I was humming. I hadn't hummed for years, and I thought, this is my body's way of telling me I'm on the right track.

At that first meeting, what we talked about was, Bill McKibben's book, Eaarth, and 350.org, but we also started asking about the major issues in our state. As a happenstance, just that week there'd been something in the paper about the Keystone XL coming to Nebraska.

The next meeting was at least double the size, and Ken Winston, who was the Sierra Club lobbyist and an attorney came. We started talking about Keystone XL and realized this was a big problem for our state. There were serious water and eminent domain issues with our ranchers and farmers, and there was some real government corruption involved. There were soil issues because they were talking about the Sand Hills, and there were real quality of life issues to have something like that coming through our state. So we decided to get to work.

The funny thing about this group is that we didn't think we had any chance of success. We decided to do it, not because we expected to succeed, but because we are all fighters. In my case, my anecdote to despair is action. If I'm feeling really badly about something, if I can work to try and make it better, I calm down internally. So I was really doing it for my own mental health… I didn't expect we would succeed. But actually, three and a half years later, we can say we're succeeding.

Let me start by saying, one of the reasons Nebraska has been the focal point of so much protest is we are literally sitting on the largest fresh water supply in this country. Forty per cent of the water used for drinking and irrigation comes from the Ogallala aquifer, almost all of which is in our state. The other reason is this is a state with very little in and out migration. It's not a beautiful state. It's a kind of a poor state, and the people who are here tend to be people whose grandparents homesteaded many years ago. We know the land. We almost all have connections to farms and to farm families. So when there's a PR blitz telling us there won't be any problem with a tar sands pipeline going through our Sand Hills, most Nebraskans know what the Sand Hills are like and how fragile that eco-system is. Most Nebraskans understand pretty well that this state depends on water, so they're very leery of any kind of tampering with our source of drinking water.

Water is our primary economic engine - for agriculture and ranching and tourism. This is a very hard state to fool.

The interesting thing was, our legislators were all pretty much bought and paid for. So it was very hard for the people of this state, who started having a lot of concerns, to get much done. But in terms of media, in this state it was really good. We had very good coverage of these energy issues, as well as the pipeline issue from our local paper, almost from the beginning. Lots of other media was quiet, but our local paper, and some of the rural local papers, really took it seriously and deemed it their responsibility to inform the public about the Keystone XL. That had a tremendous impact in this state. Readers were reading the truth about what was happening with the pipeline and the activists.

I read in Paul Hawkins book, which I think was written in the mid-eighties, that there were two million groups working worldwide to deal primarily with local environmental problems. I figure that probably means that by now there are five million groups at the rate things expand. If you think about it, five million groups, all over the world, working to change the world, is hopeful, but most people don't know that and we're not connected to each other. So people tend to reinvent the wheel in every community. I don't know exactly how this will happen, but between technology and all the extraordinary creative energy there is around these issues right now, I think within a few years we will all be connected.

Three years ago, when we started talking about this stuff, there were virtually no environmental groups in Nebraska at all, and very little movement on real critical environmental issues. There were a few people who'd worked on these issues for a very long time but they'd been crying in the wilderness. So, when we started this group, one big surprise was my original coalition was primarily urban progressives and environmentalists. In the state of Nebraska, a very red state, we never win any battles. I wrote in my book, The Green Boat, about how we're the royalty of lost causes, my friends and myself. We really didn't expect to win because we never do.

To our amazement, our first allies were ranchers and farmers. Because they need water and also they were being bullied by TransCanada and threatened with eminent domain. Ken Winston bravely went out to little VFW halls in small towns where you couldn't even admit you're a Democrat. He would go in and introduce himself as an attorney for the Sierra Club, and he would get applause. We were all on the same team. That is one of the beautiful things about the environmental movement; we're all on the same team. We all want a future for our grandchildren. We all want clean air. We all want clean water. We all want trees. We all want birds. Once you can connect people around these core issues, you're on your way. One of the things our group found that led to us being extremely popular and successful in our state, in spite of the very polarized state we live in, is that if we talk to people about the issues on which we can all agree, we could really make good connections with them. We could start things that had never happened before.

In our group, we talk about how the language we were using three years ago is now the language of our legislators. So one of the things we discussed was what kind of language we want people to be using in our state to discuss the really critical issues in the future? We decided to start using that language so that we can see it show up in 2017. In this new language, we want to be talking about things like conservation - not just energy efficiency.

The media coverage and the language used in this country can really be misleading. For example, here's a phrase I think about quite a lot: 'quality of life'. Now quality of life, the way Americans use that, is a kind of peripheral thing. Like the really important things are business and the Dow Jones and politics and corporate earnings and so on. Actually, when you think about what quality of life is, it is: water, land, animals, people, relationships, health, community, good education, enjoyment; its life. Why not just call it life? Why have some small, diminutive label for what we now see as so important in our culture?

In the beginning, our politicians paid no attention to us. More and more people were concerned about the Keystone XL, but our government was locked down and unwilling to even consider a special session to enact some rules. The rules in effect were already in TransCanada's pockets. So we decided to organize statewide so that we would have enough mass that our politicians couldn't ignore us. And we're thinking, how do we do that in this state?

It had to be positive and fun and very polite and well mannered because Nebraskans don't like radicals. There had to be no whiff of activism or protest involved. So first we declared everyone in our group a speaker, and booked speakers into any group that would have us. We did things like organize a Festival of Water on our website where we encouraged everybody to create an event locally and then post it on the site. We had over thirty events. We had people delivering wild flowers to the state Senate offices. We had a great big, "Shine a Light on Governor Heineman," night with over a thousand people walking around the mansion, singing and shining their flashlights on the governor's mansion. He didn't happen to be home that night, by the way. We had poetry readings. We had tractor pulls. We had water parties for children in the parks.

Another thing we organized was our Apple Pie Brigade. Our Governor was paying absolutely no attention to this issue, so we got a whole bunch of grandmothers involved in the cause. We started standing in front of the governor's mansion every Monday at noon with apple pies and corn muffins we had baked, and tomatoes from our gardens. And, of course, he would never see us but we'd continue to ask for an appointment.

Then one time we were there and an old neighbor of mine, who was the governor's Chief of Staff, walked into the waiting room where we were once again getting the brush off. I went over and gave him a hug and asked about his children. (That's about as calculating as I get.) I knew this man didn't really want to talk to me but I wanted to get somebody's attention and I figured he was my best shot. So I asked for his help and threatened to call him if I didn't hear from him by the next day. He looked rattled. We got a meeting an hour later with the governor.

We went in and talked to the Governor, and I think that meeting had an effect on him because, first of all, we're not radical. We're a bunch of grandmother schoolteachers who'd been bringing him apple pies. More importantly though, when the governor would try to give us some of his official bullet points around the Keystone XL, we had the information to refute those points. We asked the governor for a commitment that, if the legislature passes legislation to stop the Keystone XL, he would sign it. His answer was "I'd have to talk to my people about that." By his people he meant the Attorney General who, as I kindly pointed out, was already involved in a civil action suit for taking money from TransCanada and on record for saying he thought there was no way we could legally stop them from being involved. We said, "Governor, if you do the right thing, we will bake you an apple pie." The governor called a special session just after that meeting.

Of course, we don't take total credit for it, but we like to think the Apple Pie Brigade helped.

Usually in this state, we hardly pay any attention to our politicians. They don't do much and we let them not do much. This is not a state where there's a lot of interest in politics. So politicians weren't used to all these people showing up in their offices, and writing and calling all of a sudden, and it made them nervous.

We made a decision very early on – and this is very important for people who are trying to save the earth - that we would not deal with climate change deniers. In fact, we wouldn't even talk about that issue because that crazy issue, do you or don't you believe in climate change, has kept this country mired in a ridiculous discussion for thirty years now. John Hanson, President of the Farmer's Union said it very well, "You can't reason someone out of an opinion they didn't reason themselves into."

We kept our focus very much on the local issues. Close to home is where your heart is; it's where you are likely able to sustain commitment to causes over time, if not decades. If everybody takes responsibility for their place; that's all we really need to do.

My experience of being an environmental activist this last three years has been wonderful. I've met so many people. I've met a lot of young people I wouldn't have met. I met a lot of very engaged ranchers and farmers from rural Nebraska and a lot of passionate, ardent activists who've become my community. You know, on one level, we all know that everybody in the group would die for what we believe in, we will put our bodies in front of the diggers if they come into our state. So there's a very intense feeling of connection about the most important things. On the other hand, we have fun - we laugh a lot - we celebrate victories. Almost all of our work is about action. Every month we have new actions because we want to make good things happen; we don't waste time on what we can't make happen.

Many people, when they think about the fate of the earth, are overwhelmed with despair. Unfortunately that despair leads to apathy, numbness, innervation and passivity. A belief that you're powerless becomes a self-fulfilling prophecy. The opposite is true as well - the belief that you have power creates power. It's been very interesting to watch what can happen. One of the reasons the Keystone XL is such a good cause is, as opposed to say something like CO_2 levels in the environment, is that it takes place in real places. It crosses real rivers. It goes across real farmers' and ranchers' land. So there's three thousand miles of tangible property that we can work with as we organize people. It's a very powerful force when people can see what they are fixing.

It's very important for people, as they come to terms with reality, to be connected with other people, because other people and action with other people, generally is what gives people that sense of hope. If they try to act alone, they're very vulnerable to despair. If they act with others, they're likely to experience what Durkheim called 'collective effervescence,' which is a certain joy that comes from working with other deeply committed people on a cause you know is just. Martin Luther King called that 'beloved of community.' And that is a really important thing.

Leaders are starting to emerge everywhere. Nebraska is a good example of that. Three years ago in this state, we had no environmental leadership. Now environmental leadership is scattered across 75 groups, 100 to 150 people. We can barely keep up with the number of groups and actions going on in our state. Once a few leaders emerged, the cause is so ripe for action that everyone wanted to be a part of it.

So I'm actually pretty hopeful. I mean, no one really knows what time it is right now. We could almost be at the end of the human era … or we could be at the beginning of a new and marvelous world, not only for humans, but also for all living things. And once there is a change in consciousness, things can happen very fast. It won't be a long political struggle; it will happen overnight."

THE HAND OF GRACE...AGAIN

By our last day in Nebraska, I had interviewed at least 20 or 25 people about the fight to stop the Keystone. I knew I had enough interviews. What I did not have was what filmmakers call B-roll - background shots of the land that can be on the screen when someone is talking, so the viewer can see each region, not just hear about it. I asked Chelsea to cancel the rest of the appointments on our last day, so that we could head out to the country and get b-roll shots.

Divine Grace had a different plan. As we got into the car, and started out on our search for beauty shots of the Sandhills, the weather got more and more insistent that there would be no outside video taken today. First it began to rain, then the rain turned icy, then the dropping ice turned to snow, and to top it off, it even got darker as fog rolled in. 'Stop,' I finally yelled, laughing, as we headed back to the house in Omaha where about six canceled interviews were being rescheduled.

When we arrived in the historical district of Omaha, people were waiting to be interviewed. There was only one person in the house that was not scheduled for an interview, and that was the homeowner Steven Evans. Later he would tell me that normally he would have been upstairs working in his office, but he just had a feeling that 'perhaps' he should stay down stairs. We ended up getting a great interview with him, but it wasn't about the Keystone – it was about Fukushima.

"Hello, my name is Steven Evans;" he began, "I'm both President and Senior Research Scientist for the Therapeutics Research Institute, a nonprofit 501(c) 3 corporation, here in Omaha. Our mission is to aggregate and pull together the medical wisdom, insights that have been developed worldwide, and put them together in comprehensive protocols, or recipes, that people can implement by themselves to enhance their own health status. Based on what has already been discovered, but is not yet part of the standard of care of American medicine, Some of these standards of care are well known in other countries already, so it's not necessarily wild and crazy whatsoever. We sell no products, and we have no charges; we simply make information available as a nonprofit to aid people in taking the initiate for their own healthcare.

On March 11, 2011 a tsunami hit the nuclear power plant run by TEPCO, and as a result of the earthquake, the power plant, which has multiple stations, was brought to its knees. In spite of repeated denials, there's now an admission by TEPCO that there's been total and complete meltdowns of a variety of those nuclear stations at the plant. As a result of one of the meltdowns, I believe it was station number three that literally blew up and blew the roof off. Primarily to save money and based on the design, that particular component, that nuclear generator, had about 1550 rods that are perhaps six, eight, twelve feet long, and they were stacked in a container 100 feet above the ground. They are there, they are now exposed and continuously being cooled, they may be impaired and they represent what constitutes 1550 atomic bombs waiting to go off.

After a couple of years of thinking about it, TEPCO's desperation has driven them to attempt to remove those rods because the building that is housing them is in a pool of water, a hundred feet above ground - not a very clever design I might note - that is leaking. The building itself is sinking into the ground unevenly because the radiation's intensity has actually degraded the dirt itself, and this constitutes another potentially devastating explosive prospect. If there is another earthquake, another tsunami, you will have the level of explosion and release of deadly cesium-137 that could clearly fry Tokyo. It represents an enormous danger.

Besides local earthquakes, which affect Japan all the time, there is the possibility of another tsunami hitting the power plant again, or an eruption of Mt. Fuji that could bring that power plant generator down. In the worst-case scenario you would have a level of explosion that dwarfs the Chernobyl explosion that Russia had to contend with quite a number of years ago.

There are apps on the Web now for those who want to be tantalized by this. They give a continuous readout of the Island of Japan and the ongoing earthquakes that are affecting the island regularly. As I first looked at the app, a 7.2 earthquake, enough to bring down this plant, occurred several hundred miles north, and not too long ago one occurred in the ocean 200 miles due east. So you have Nature playing Russian roulette.

TEPCO finally decided they had to start removing the tubes, but the placement of the tubes previously were absolutely precise, guided by a computer arm, within exact millimeters of each placement. Each of those rods now have to be pulled out without touching any other rod, and they have to be extracted absolutely linearly because if they hit the sides and break or touch another rod you could have an explosion. Without the computer arm - which was blown away – and without the precise control, TEPCO is attempting to do it by human hands. They are eyeballing the rods, pulling each of them out with a precision that heretofore has never actually ever been achieved by mortal man, so the prospects for disaster are simply enormous.

In the meantime, recently gasses have been seen being emitted from the plant. The speculation is that it might just be steam, but in reality, it could actually be a nuclear catalyst occurring because the total meltdown has created a completely open nuclear event - quite an extraordinary situation. So you have the potential for a catastrophe that is simply world class. It's estimated that the cesium cloud would rise about ten miles up, and be caught by the jet stream and distributed worldwide … within 72 hours. Respected nuclear physicists from Japan have said that it would fry the west coast of California at the onset.

The event is extraordinary, the threat is enormous, and so the question would be why haven't I heard about this? Why did I just hear on the news that it's all being taken care of? The answer is actually, in my opinion, very simple, because if there is nothing to be done, why instill worldwide panic? Why have the recognition of a pandemic if there is nothing you can do about it and no way to make money from it?

There were a number of remedies for human radiation exposure that have been uncovered which are tremendously effective, but the very least of which is iodine. You may read that one can just take iodine if there's a problem, but that turns out to be blatantly, medically false. Iodine will bind with the thyroid, which is iodine hungry, which means you could protect the part of your thyroid it binds to, but that's it. Radiation is not only going to hunt your thyroid [chuckle]; there's your brain, your heart, your liver and the rest of your body.

After TEPCO experienced the tsunami, earthquake and explosion, The Japanese Research Institute in Japan began researching what to do for radiation employees who had already been direly exposed, and they came up with a very simple solution. They took 25,000 employees and treated them with high-powered intense IV solutions of vitamin C. The other group was simply left to their own devices. Ninety days later, there was no measurable radiation damage in the treated group. They could detect no DNA breakage - they had no inflammatory processes that get kicked off when you get radiation. The other group was essentially near death. The Institute reported all this to the Japanese government, in an official report and the Japanese government rejected the report.

Why would they reject that report? Because they had no problem, and if you have no problem then you don't need a solution do you? The Institute sent it to Korea and to Singapore and the report was rejected by each of those governments and returned. The people in the Institute, who are MDs, PhDs, and all respected researchers, leaked that report to a number of journals, none of which would publish it except one; The Journal of Orthomolecular Medicine. That published peer review literature said there are a number of things to do, so we found their literature and we put it together.

In fact, our own research went one step further. Once you know that high level IV vitamin C is remedial for explicit radiation damage, you find out that it's pretty hard to get and it'll be somewhat expensive. So we found a supplement that allows intense Vitamin C to be absorbed as effectively as an IV. You might rush to Walgreens and buy Vitamin C but that won't do you any good because after you take a number of capsules the gut will no longer absorb it and the result will be diarrhea. You will need a special kind of C, and it's called **Liposomal C**. Various companies on the Web sell it. All the information is on our website. http://www.wikicuria.com

So there are solutions, they're known and available, and again I believe that the heart of what we consider to be a cover-up is related to the unreasonable strategy of why panic everyone? It's not good for economic growth. However, you can, in fact, purchase these things by yourself, they're relatively inexpensive, and you will be able to take preventative action to protect yourself or even use it to reverse dire exposure. I search the Web for medical information for a living. Nowhere can I find anybody undertaking the effort to make this information available.

There's another interesting question when it comes to the Fukushima cover-up, having to do with the fact that Tokyo would be hosting the 2020 Olympics. Can you imagine a more devastating PR event than to say, 'come to Tokyo and by the way, you can take back enough radiation to glow in the dark?' Not an optimal PR event but at least you would save electricity (chuckle) at home. It is so counterproductive to even recognize the problem that the government wisely, for economic reasons, should be in complete denial. But for health and moral reasons it's an absurdity.

We have associates in Japan with Geiger counters that show there are hot spots in Tokyo now that are direly dangerous. Places that cause the Geiger counters to go crazy, but since Japan is so homogeneous in its cultural nature, dissent is suppressed far greater than anything here in the US, and the PR would be unbelievably devastating.

On the west coast of Canada, for two weeks after the explosion in Fukushima and the release of cesium gas, our own EPA ran tests. Strangely enough, the levels were only modestly rising. They declared there to be no problem and stopped evaluating it. Another person might ask, slightly more cynically, what if we keep measuring it as that cloud moves across the ocean, hitting Hawaii and the west coast, what about the subsequent impact? Our EPA said, "Well we have no problem right now, and therefore we don't have to do any further testing."

Individuals, researchers like ourselves, have set up monitoring groups and given people a little dosimeter card, like healthcare workers would wear, to keep in their house to see if there's any increase. Repeatedly there have been extraordinarily high levels of cesium found. Fifteen great white tuna were captured; all had 15 to 20 times the previous levels of cesium recorded. You'd think that might create a problem. Their response is, "It's nothing to be worried about."

If you think about it economically, you don't want to indict the fishing industry off the west coast, and so you get continuous contention between powerful economic interests and the actual health of the people. And you also have the Canadian government which published reports that said they have no measured cesium. What's going on now on the west coast and other areas, as the level of the background radiation increases, is that officials are also raising the background level. So if 'normal' levels are raised, can we tell if we have more radiation? Well, the answer is a very convenient no. If you keep cranking up the background level measurement, then everything continues to appear normal.

Not only did Chernobyl spread around the world, but also we have had other kinds of events that emit radioactivity. There has been a certain level of 'background' radiation that the population has experienced for decades, even centuries, but now that the 'acceptable' level has statistically been raised by the government. Despite all this cover up, the facts remain that radiation is rising. But the system cannot afford to recognize this problem, so it redefines what's happening so that there is no problem.

In Japan they took a samplings of children, and not just in Fukushima, and checked their thyroid. They found 32% had thyroid nodules, which is precursor for thyroid cancer. This is an extraordinary level; and in fact it is unheard of. So recently there was a published release by the Pediatric Association in Japan, that thyroid nodules will no longer be reported. Parents were apprised that there's nothing to worry about. Now I assure you, if you had a child in this country with thyroid nodules they would be monitored intensely, and probably over treated. Children, some experts have said, are about a thousand times more eligible for radiation damage than older adults and therefore need greater protection.

As we look to the future, what can we do to short-circuit this impending disaster? What's the technology, what's available, what interventions can happen? Sadly, it is generally recognized by experts that there is not yet any known technological, methodological approach to a situation that has never occurred in the history of mankind. This is a complete and total meltdown.

If you look at some of the webcams of the Fukushima plant, you'll see what looks like a mass of Jell-O - an entire disintegration of radioactivity, uncontained, and totally broken through all containment and concrete receptacles. Such an event has never occurred, and currently this destruction is sinking into the earth. The fear is that once it hits ground water it will then cause an entire cataclysmic eruption. Has that ever occurred in the history of mankind? No, it never has.

So what expert opinion has said has been we need to create an international consortium of experts to bring them to this problem rather than rely on a local power company in Japan to try to produce an answer. Why hasn't that happened when that call for international intervention has been repeatedly given and when the US has offered experts and Russia has offered experts from their Chernobyl experience? The answer from Japan is the equivalent of "I'm okay Jack, [chuckle], I got it under control, there's no worry, and it's okay." Culturally incorporating outside expertise is incongruent with their cultural nature. They have rejected interventions from the US and all others.

Just recently, under enormous pressure, a panel was created to review TEPCO's efforts. That panel released a report that said we are absolutely aghast by the situation, by the calamity, and by how little expertise is available to address it. All well and good, but it has not significantly increased international intervention. Then let's return to the question, what could international intervention provide? And the answer, again, is that there are no known solutions to a problem that has never before existed.

There have been a number of groups in the world, including what are called the Elder Brothers, who are high in the Andes, who came down to the lower villages, which was quite remarkable, where there was a press conference. This is a translation of their warning:

> "We are the Elder Brothers, we have been here for untold time, and we've come to bring the message the Earth is in dire danger. The Great Mother taught us what we needed to do, and her teaching has not been forgotten right up to this day. But now they are taking out our Mother's heart, they are digging up the ground and cutting out her liver and her guts. The Mother is being cut to pieces and stripped of everything. The great Mother too has a mouth, eyes, and ears. They are cutting out her eyes and ears. If we lose an eye we would be sad and the Mother too is sad and she will end, and the world will end, if we do not stop digging and digging."

That call came forth from the Brothers, I believe in the year 2000, and they recently came back down again, the second time in recorded history that this group has deigned to talk to the rest of the people of Earth. This ties to such things as the XL Pipeline and to all kinds of climatic changes that we are creating. They said that we desperately need to change our thinking, need a transformation of our moral perspective because, we are so close now to the tipping point that without a revision of our deepest thinking we will have doomed this planet.

Many of the famous world's physicists and nuclear experts say that Fukushima threatens humanity in its deepest way ever. And so I choose to think of Fukushima as the catalyst, the call for necessitating a change of consciousness, a change of what we're about. We must rethink what we're doing. But of course even that doesn't ensure success, it just ensures that we see the challenge that is right in front of us."

Steven Evans was my last interview in Nebraska. And to think I almost chickened out because of the cold. In those five days, I had discovered people who had stood up to political corruption and big businesses, not because they thought they could win, but because it was the right thing to do. I had made friends that hopefully will last a lifetime, or for however much time we have left.

By the way, while we were filming there, a lot of people laughed at me. Even the people I interviewed doubted the wisdom of my extensive coverage. Many said 'Thanks for coming out, but I am afraid your film is going to be obsolete when Obama approves the pipeline." At which point a voice from deep inside of me would reply, "Obama is not going to approve the pipeline."

As it turned out, almost two years later, the day before our sales agent from American Cinema International, Chevonne O'Shaughnessy, took the film into the market place, President Barack Obama rejected the Keystone XL pipeline. That kind of Devine Timing is both humbling, and triumphant. We Know Not What We Do was the only film at the market that had a camera on the ground recording that XL Pipeline history. Who could have predicted such timing?

THE EPILOGUE IN THE EDIT BAY

After Nebraska, it was pretty much wrap time for 'We Know Not What We Do'. As we started the tedious work in the edit bay, looking at all the wild footage and interviews, it became apparent that we would have to return to Nebraska for that b-roll the weather had denied us that last day. What better time to do it than June when Nebraska would be in full bloom? My editor, Grant Taylor, is also a great cameraman, and so we headed back to Nebraska…just the two of us, carrying a camera.

This time, Art Tanderup was our guide. Art was insistence that we needed to film 'Spirit Camp.' We learned that Native Americans from Rosebud Reservation in South Dakota had established that encampment as their resistance to the Keystone. They had started Spirit Camp, in the dead of winter, to block the route of the XL pipeline.

After my visit to Nebraska in January, I had great respect for any group executing an outdoor project in the winter. The practical part of me felt we didn't need more footage, but the storyteller knew that if we were being invited to Spirit Camp we needed to go.

As it turned out, 'Spirit Camp' was eight teepees pitched inside a barricade of stacked wheat, but the feeling of peace, and other world wisdom, made it feel much bigger and more protected. There was a huge tarp tent that served as a soup kitchen and gathering place. It had all been placed on land that the pipeline would need to cross before it could reach the Sandhills. The tribe had announced to both TransCanada, and our own federal government, that Spirit Camp had been built to block any attempt of the pipeline going through. They added that any physical attempt of the pipeline going through that small plot of borderland would be a declaration of war.

A resident of the new village, Chris Fire Thunder, an Oglala Sioux, showed us around. He shared his love for the camp, and let us know that the residents and guests pray for the Earth Mother daily, thanking her for the privilege of living on her womb. In fact, Chris said, "We say prayers for everyone, including the Koch Brothers and their families. We don't have no animosity; 'they have just gotten a little off track in their journey."

During that interview, Art got a text inviting us to drive on to Rosebud Reservation headquarters where a pipeline meeting was in progress. When we arrived, I was leery about walking in with a camera, so we hung back in the parking lot to get our bearings. Almost immediately, we were told to bring the camera and come inside. Once inside, Grant wisely started setting up an interview corner even though we didn't yet know whom we were interviewing.

Here then are the last two interviews that we gathered for our film. The first is with Charles Marshall, who shared the Lakota Elders vision that dates back one hundred and sixty years. The other is with the great Spiritual Chief, Leonard Crow Dog.

Charles Marshall
Coordinator of Spirit Camp
Shielding the People

"My name is Charles Marshall," he began, "I am the Coordinator for the Oyate Wahacanka Woecun, which is the 'Shielding the People Project'. Our Spirit Camp, established on the east boundary of our tribal land, is playing a big role in how we see ourselves now. Not only as a tribe, but how we are choosing to identify ourselves as the Oceti Sakowin, of the Seven Council Fires. The tribes are coming back together, because of this pipeline thing. On several occasions our Rosebud tribe has met with the other tribes that are on board with the opposition of this pipeline. At times we have even, in a sarcastic way, wanted to thank KXL Pipeline for bringing us back together, for bringing this unity, and for allowing us, for the first time to look at ourselves as a sovereign nation again.

We've been talking about sovereignty for a long time. This cause - and our passion to stand up for our rights – this unity - is built around people coming together. For the first time tribally, all the seven tribes and reservations in South Dakota, Dakota, Nakota and the Lakota are back together. We are walking the walk, instead of just talking about it for the last 50 years.

Most people don't know it, but this pipeline was prophesized 160 years ago. The Elders called it the 'Black Snake'. They weren't quite sure what it was that would bring the people together, but they identified it as a black snake that would awaken the people and pose a need for defense. They said we would have to defend ourselves, our families, and the water rights, and the land, and our four legged relatives, and the winged.

So we are a part of history, of something that was prophesized long ago, and to be partaking in this is very humbling. We are the answer to the prayers that our elders prayed hundreds of years ago. The fact is that since their vision, we have been through many challenges, even the oppression of genocide. I look at myself, as I'm sure a lot of Lakota, Dakota, Nakota's look at themselves, as a miracle. Genocide isn't something that a lot of people; or a lot of nations, make it through. The Lakota people, the tribes, are still here today and we are coming back to the point where we identify ourselves, through our ceremonies, through our language, through our songs and through our history. Being a part of this movement, going with the momentum that has been built up because of this pipeline, to me is being able to see true hope for the first time for our people. It's having faith in our prayers.

When all of this started, the more I talked to my tribal elders about our involvement with the pipeline, the more interested I got. I was very honored when they asked me to come aboard as the coordinator for this project. I get to be a part of the whole project, to form and establish an infrastructure, to maintain it and build a community out here. That pipeline, the Keystone, is hanging on by a thread, and we're praying that it's going to fold under its own weight. Seeing that vision, sharing both the visions of today and long ago, we have come together to build a dream and in my opinion, to complete our prayers, and to answer the prayers of our ancestors.

This project, in my opinion, is something to be shared and now that the pipeline is going to be on hold for a couple of years, we are talking about building a real infrastructure out there, a permanent, infrastructure, with classrooms, and a museum, and a cultural building that would teach our own people, our youth, about our spiritual rights, our Star Knowledge.

Our Star Knowledge lessons teach us about our very existence, not just here on earth, but how everything here is connected to the entire universe, how our actions are felt within the entire universe. My involvement began with little knowledge, a lot of compassion, a lot of passion about what I wanted to know, a lot of interest, and it's been such a honor to be a part of this, to be a part of the prophecy, to be a part of these prayers, these dreams, and to move forward, to see real progress in light of where we've been, and how we've gotten here today. To see hope in each other.

I had an elder tell me one time about his visions, and when he was finished he said the strongest words that I'd heard in a long time, "we're coming back now." Those words have stayed with me ever since. When I heard them it gave me goose bumps, and a breath of fresh air. I felt the drive to know more about my own people, my own culture, my own ceremonies,

The history of our people has always been oral, our most important events, were always drawn on a buffalo hide. Symbols are a strong part of our culture as they connect us again to the stars. I know that there is a drawing, an old drawing, that

I would like to find, of the actual vision that was given to the Elders about the Black Snake.

I don't know if we can know the outcome. I think that is determined by the prayers that we are making today. And for me, to hear the prayers, to be in ceremony, to see the commitment from the people that are involved, and to see for the first time in, in my life and I'm sure in my elders' lives, to see for the first time Cowboys and Indians coming together as well because of this pipeline. Seeing our connections, seeing our sameness, not our differences anymore, that is the most powerful thing that has ever happened to me.

Our project is a big part of educating people, as well as educating ourselves about our involvement environmentally and the impact that it's having. The alliances that we have established in our Shielding the People project has gained recognition throughout the world. We have alliances in Australia, people who are supporting us in India, and so to me, being a part of this coming together is everything. Our words that we choose to use are important, how we conduct ourselves is important, and I couldn't have asked for a better time in my life, a better purpose in my life.

Personally speaking I've never really understood activism before. I've always been around people that have been activists, but never in my life have I ever imagined I would have the opportunity to stand in the center, with my people, and send out a message that we all share, about the one thing that connects us all. Water. We all need to drink water, and this oil is threatening that; it's threatening the rights of every individual, whether you're Native, whether you're non-Native, it's threatening clean drinking water, clean air, a clean environment, and for the first time I think a lot of people are being more environmentally aware of what is actually happening. Nothing on this planet can exist without water."

Charles Marshall continued to talk to me about the First Star Knowledge Conference that happened in the nineties. As he spoke, the last twenty-five years of my life began to make sense. Oddly enough, I had attended that first Star Knowledge Conference that was held in a high school in South Dakota in 1993. Very few people have ever even heard of it, and yet here it was reentering my life, 20 years later.

The Star Knowledge conference had brought together Lakota's leaders and UFO experts, because the Lakota's were saying that it was time to reveal ancient prophecy to save our Mother, the Earth. Strangely, I attended, armed only with a tape recorder, and was fascinated by the interviews, and being a part of it all.

Out of that encounter had come an early morning phone call that reached me back in Los Angeles. It was a Lakota Elder, Floyd "Looks For Buffalo" Hand, who told me the Star Children, who are circling the Earth, had contacted him. He was told to instruct me to bring a film crew to Yellowstone National Park, to help save the last of the wild Buffalo. From that amazing call, I managed to make my first documentary 'When Buffalo Roam', in a desperate attempt to stop the slaughter of that last wild herd that roamed free in Yellowstone park.

I had never even seen a Buffalo in this lifetime, but I was so impressed with the imagination of the call that, without any money, I began work on a documentary that very day. I learned that the small gallant herd at Yellowstone was the remnant of the wild herds that once roamed the Great Plains, sixty million strong.

My short film, When Buffalo Roam, won the 1999 New York International Film and

Video Festival. And even though it was only 9 mins long (edited out of 40 hours of footage), it brought a lot of attention to the plight of that herd who were being both slaughtered and harassed by the Montana Dept. of Livestock because of the greed of the cattle industry. (www.indievue.com)

When all of that was happening, I had wondered how the Native Americans had found me, how did they know of my film experience, or my yearning to direct? It had never quite made sense ... any more than the sudden invite today to bring my camera to Rosebud. But Rosalie Little Thunder, whom I had reached out to after the call from Floyd Hand, had been from Rosebud Reservation. And getting to know beautiful Rosalie had begun this entire phase of my life, leading all the way to my current fight to save our Mother. As I listened to Charles, all those mysterious dots that had made up my life began to connect. I could hardly breathe with the excitement of it all.

After we finished that interview, the man I thought would be our next interview came to me and announced, "You are now being blessed with an interview with our most spiritual leader." I have been around Native Americans enough to know they have a very subtle but wicked sense of humor, so I honestly didn't know if he was teasing me, I only knew that I dared not laugh. What happened next was an incredible blessing, both for me, and the film.

Chief Leonard Crow Dog, the man who has prophesied "human beings have only a few more years to stop tearing up Mother Earth, or she will take herself back from us," is a fourth generation Sicangu Lakota medicine man who became well known during the second takeover of Wounded Knee on the Pine Ridge Indian Reservation in South Dakota in 1973. He has sought to unify all nations of Native people. He is also dedicated to keeping Lakota traditions alive. Chief Crow Dog speaks five different

Indian languages, and though his English is broken, his words, and indeed his very presence, penetrated my soul. He entered the room quietly, wearing simple clothes, and took a seat.

Chief Leonard Crow Dog
Spiritual leader
Shielding The People

"Good afternoon Relatives, in this year of 2014 in the times of the life we live," he began. "My name is Crow Dog, and I am the spiritual overseer of 562 tribes. What I want to tell you today is that we all live on the minerals of the Dural-thermal power of the Universe, our Grandmother Earth. I have worked with these facts for many years. Now we have come again to be a unified family, and to bring this land back for the better.

There are those in the biggest corporations that do not understand the land, or the tree of life or the people who live on the land. But if you look, it's still all there; the buffalo, the deer, the fish, and all the stock that live naturally. They do not assimilate changes.

We brought a salmon from northwest to check our water; he was very comfortable. Same as the beaver, we brought him from northwest, and a fish from the center of the ocean. The sound of communication is always with us, and we have communicated with them real nice; it's a beautiful way of life. So our only difficult challenge is the governmental purpose, the Department of the Interior; and that path is very hard for us. We have to talk with their attorney on their terms. We have to have an attorney to help us put words on paper to justify our theories.

We want to explain our theories, about how to better live on this Earth. We want to put our theories into an understanding. As we proceed, the word we all must learn is 'excuse me.' When we put the tobacco in the pipe to pray, when we test the sacred pipe to make it better and make it good, we need to say to each other, excuse me. It's all going to take some time, but it's possible.

I speak five Indian languages besides my own and what little English I have is broken, but they say I have a clear understanding. We want you to know these things, so that you will have that clear understanding too. Together we can verify it, so that it's clear for future generations. We know the afterbirth of this child that is being born. We know grandma, we know grandpa; we know the pain the same as the land, but it is all worth it because the land is holy to us.

There are many treaties that counted us as citizens of the United States. Then there was the 1934 Reorganization Act where we lost five regions given to us in those treaties. But despite all this pain, all the lost, there's some elements of our Mother that are still speaking to us, and those elements know that we are with them now. It's all very holy and sacred to bring forth this unified family, this Dural-thermal power of the universe, this telepathic communication that the world has never known. It is old and from our family Tree of Life.

And so we know where it's at - the water - we don't own the land; the land owns us. But She needs to be protected. She has gone through cesarean; she has gone through a lot. The highest human intelligence, doctors, professors, even the attorneys feel that we own the land. But according to the original instruction of the Supreme Law, we are only the caretakers.

So those elders before us, that called it a Black Snake, have to be honored. We must not lie to them. We've been lying to them for many years, now we must come together to make it good for the future. So how are we going to bring us together for this joint work? By bringing the understanding of this way of life. We are bringing this child to be born again, to know the language and to know the concern. Pavement and concrete are not the answer, the answer is the divine roots of this land, and we are part of that.

We are the living Buffalo Party. We know when to drink the water; we know when to eat the buffalo grass. So if I say buffalo is standing outside, you may look outside and say there's no buffalo, but there is buffalo, there is buffalo grass. The oldest nutrition we have, and it has to be protected because it's holy, it is sacred.

You look around, there are not too many elders now, but the elders shall be born again. And that's what it's all about, this surface of the universe. The surface has to be continued, we have to tell that to our kids, and our kids will tell their moms and dads. It's not alcohol that answers, it's not drugs that answers, man's creation is good, but it misconstrues. Ask the highest educator; we already have a universe of the university. In tablets it is written and in eons and generations, it is our way of life. It's not something new; it's something holy. And it's going to be a good, a beautiful understanding. In the word of the American system we will say 'I'm sorry.' And all of us need to learn to say 'excuse me.'

We're not by nature a greedy people. We have lived a life of history. Our way of life is very natural. We made this history, and now we're going to follow through. We must all learn to live in the web of life. And that's it."

I sat in stunned silence for a moment, then I said to him, "Chief Crow Dog, I am very moved by your words, I too want us to work together, all people working together as children of the land. So even if you say to me, there is Buffalo outside this window, and I look out and don't see a buffalo, I know that the buffalo is still there."

And as his twinkling eyes looked into mine, he smiled his wonderful smile and replied, "That's it," as he quietly rose and walked away.

That's It!

There is one more thing I must add before we close. I interviewed Leonard Crow Dog in June of 2014. In deciding to do this book, I went back to the transcripts and wrestled with his interview, it was hard to understand in his broken English. I had been so deeply moved by meeting him. I had the feeling that I understood what he was saying, and I wanted you, the reader to have the same experience. So as I worked with his transcript, I silently asked for his help in interpreting his words. That night I had this dream:

It was the Sixth Stage of Extinction…

All the animals were disappearing around me. Small animals, bigger animals, were all disappearing from the surface of the earth, until finally, I was the only living creature left. A terrifying giant yellow snake was now coming rapidly toward me. I was about to be bitten and/or swallowed! I grabbed the snake by the throat and began a life and death struggle. Tiring, and realizing that the snake was gaining ground, and that I would soon be gone, I began to shout, frantically, for help. "Mother! Mother!" I cried desperately.

I awoke from the dream in a sweat, so frightened that when my tiny dog moved on the bed, I jumped, heart pounding. It was all so real! Later, when I went back to work on Chief Crow Dog's interview, it had changed itself into a much more understandable verbiage, and I realized that my new shaman, Chief Leonard Crow Dog, had heard my call for help, and had indeed helped me.

It wasn't until later in the morning, when I learned that the snake, in Native American symbolism, stands for disappearing and death, that the strong medicine of the dream also became clear.

Please, be aware that we are indeed entering the 6th stage of extinction. A great many living creatures, both big and small, are starting to disappear. What is not as evident is that our own species is in grave peril. As I was about to die, bitten and swallowed by the huge snake, I kept yelling for help, frantically screaming to Mother. After waking I had wondered why I had called to my mother for help, thinking that my human mother would have been no help at all.

As Chief Crow Dog has prophesied, and as our Elder Brothers of the Andes also tell us, 'our Mother is taking herself back. We will surely perish if we do not stop digging and digging,' if we do not shift our values and realize what is really important to our lives. Staying with my dream, I suddenly understood that both the making of the film and the writing of this book has been my way of yelling to the Mother for mercy.

My prayer is that the film and this book will encourage you to be willing to make a moral and spiritual shift in your thinking - if you want the human race to survive. And if you are asking yourself, and me, what does that mean, in simple terms it means learn to love. Learn to love yourself, and then learn to love others; learn to love all things. Let that love lead to gratitude. Let that love and gratitude teach you how we can live in the bio diverse web of life.

Please realize, that despite all our flights of fancy, we are just tiny beings in a space so enormous that we can't even imagine it. If we do not make the shift to love, if we keep pretending that the Creator doesn't exist, the snake will simply swallow us and we will be gone.

So I say with love to Grandmother Earth, and also to you,

"Forgive us; for we know not what we do."